Gene Network Inference

random forests, and Bayesian network structure learning. We detail how they can produce w_{kl}^{m} and w_{kl}^{e} scores. Then, in Sect. 2.2.6, we show how one can produce a final network prediction in the form of a ranked list of edges.

2.2.2 Bootstrapping

Bootstrapping (Efron 1981) is a resampling technique for assigning measures of accuracy to sample estimates. It has the advantage of allowing the estimation of the sampling distribution of virtually any statistic using only a simple resampling approach, at the cost of repeated computations.

The general idea of bootstrapping is to randomly draw N_{boot} replicate datasets with the same sample size as the original data. Each of these replicate dataset is obtained by randomly sampling with replacement from the original sample. For each replicate dataset, the model is fitted, and it is then possible to study the statistical properties of the distribution of the considered statistic on all resampled datasets.

In this chapter, the major use of bootstrapping is to contribute to the construction of the so-called "confidence score" of edges in the predicted GRN. However, bootstrapping offers further opportunities. Since bootstrap datasets are obtained by sampling with replacement, each of them is deprived from around $1 - 0.632 = 36.8\,\%$ of the original samples (Efron and Tibshirani 1997). It is possible to use these out-of-bootstrap samples to study the behavior of any given loss function on those samples. This feature is used internally in the random forest approach and allows to avoid overfitting.

A major drawback of bootstrapping is that it roughly multiplies the computational burden by N_{boot}. The loss of 36.8 % of the data in every bootstrap sample may also affect the sharpness of estimates on every resampled dataset.

2.2.3 Penalized Linear Regression Approaches

A natural approach to solve the network inference problem is to consider each gene g individually from the others and consider that its expression value can be represented as a linear function of all other gene expression levels and of all polymorphisms:

$$E_g = \sum_{j=1}^{p} \alpha_{gj} M_j + \sum_{\substack{j=1 \\ j \neq g}}^{p} \beta_{gj} E_j + \varepsilon_g,$$

where α_g is the p-vector of linear effects of polymorphisms on E_g, β_g is the p-vector of linear effects of other expression levels on E_g (we assume $\beta_{gg} = 0$), and ε_g is a Gaussian residual error term.

Parameters α_g and β_g can be estimated for each gene g from matrices **m** and **e** using a variety of linear regression methods. The main strength of this model lies in its simplicity, leading to only $2p$ parameters to estimate for E_g, a very desirable property for estimation in high-dimensionality settings. More complex linear models including interactions terms between polymorphisms and expression levels would immediately lead to a quadratic number of parameters.

Given the fact that $n \ll p$ and that regulation networks are expected to be sparse, penalized regression methods leading to variable selection such as the Lasso regression (Tibshirani 1996) or the Dantzig selector (Candès and Tao 2007) were chosen to perform the regression.

2.2.3.1 Lasso Penalized Regression

In the linear regression problem, a response variable Y is a linear combination of r regressors $X = (X_1, \ldots X_r)$ and Gaussian noise ε:

$$Y = X\theta + \varepsilon$$

Having observed Y and X on a sample of size n and assuming Gaussian distributions, the estimation $\hat{\theta}$ of the parameters θ is obtained by minimizing the residual sum of squares (RSS):

$$\hat{\theta}^{\mathrm{rss}} = \arg\min_{\theta} \sum_{i=1}^{n} (y_i - \sum_{j=1}^{r} x_{ij}\theta_j)^2 = \arg\min_{\theta} \mid\mid Y - X\theta \mid\mid_{\ell_2}^2,$$

where y_i and x_{ij} are the observed values of Y and X_j for the ith individual.

The Lasso regression (Tibshirani 1996) penalizes this RSS criteria by the sum of the absolute values of the parameters (their ℓ_1 norm):

$$\hat{\theta}^{\mathrm{lasso}} = \arg\min_{\theta} \mid\mid Y - X\theta \mid\mid_{\ell_2}^2 + \lambda \mid\mid \theta \mid\mid_{\ell_1} \tag{2.1}$$

This penalization leads to a shrinkage of parameter estimation. More importantly, the shape of the ℓ_1-norm specifically favors the estimation of zero values, leading to a natural variable selection behavior. Shrinkage and selection levels are controlled by the magnitude of the penalty term λ. The Lasso criterion can also be written in its dual form (by Lagrangian transform), which makes the constraint on parameters more explicit:

$$\hat{\theta}^{\mathrm{lasso}} = \arg\min_{\theta} \mid\mid Y - X\theta \mid\mid_{\ell_2}, \text{ subject to } \mid\mid \theta \mid\mid_{\ell_1} \leq t \tag{2.2}$$

In Eq. (2.1), the larger λ is, the greater the amount of shrinkage, and the more parsimonious the selected model will be. More precisely, λ is an upper bound on the

2.2.4 Random Forests

In the previous section, we reduced the GRN learning procedure to a gene-by-gene linear regression problem. However, nonlinear regression methods can also be considered, assuming that the expression level of E_g is a function of the remaining expression levels E^{-g} and of allelic states M:

$$E_g = f_g(\mathbf{M}, \mathbf{E}^{-g})$$

The use of random forest for GRN reconstruction from expression data alone has been originally proposed in GENIE3 (Huynh-Thu et al. 2010). Indeed, one way to solve such a nonlinear regression problem between a response variable Y and regressors X is to try to recursively split the observed data with binary tests based each on a single regressor variable, trying to reduce as much as possible the variance of the response variable in the resulting subsets of samples. Each test is a node in a binary tree and typically compares the input variable value with a threshold which is determined during the tree growing. Ultimately, the leaves of the tree give the predicted numerical value for the response variable.

A random forest (Breiman 2001) is a collection of such trees grown partially at random, using two sources of randomness:

- Each tree is grown using a random bootstrapped sample of the data.
- The variable used at each split node is selected only from a random subset of all variables (typically of a fixed size K).

The random forest predicted response for a sample is the mean of all the regressions predicted by each tree. Besides the possible nonlinearity of the response, a specific strength of random forests lies in the fact that they can use the internal bootstrapping to estimate the importance of any regressor. After shuffling the values of the regressor considered in the samples that have not been used in each bootstrapped sub-sample, it is possible to compute the resulting increase in the variance of the regression error compared to non-permuted samples. This provides an evaluation of the regressor importance.

In the context of GRN learning for any gene $g \in \{1, \ldots, p\}$, the random forest method provides us with importance factors f^m_{ig} with $i \in \{1, \ldots, p\}$ and f^e_{jg} with $j \in \{1, \ldots, p\}, j \neq g$, that respectively give the importance of allelic state M_i and of expression level E_j to predict E_g. These weights can then be normalized for each gene (Huynh-Thu et al. 2010). For each $g \in \{1, \ldots, p\}$ independently, we normalized the f^m_{ig} and f^e_{jg} by their estimated standard deviation. All these importance factors can then be sorted producing global ranks r^m_{ig} and r^e_{ig}. The marker-based "confidence scores" for oriented edge $k \to l$ are then defined as:

$$w^m_{kl} = 1 - \frac{r^m_{kl} - 1}{N},$$

where N is the largest overall rank. A similar definition is used for the w^e_{kl}.

The computation was performed using the `randomForest` R package (Liaw and Wiener 2002). The number of trees was set to 1,000 and other parameters (defining individual tree depth or the number of variable to draw at each split for example) were kept at their default value.

2.2.5 Bayesian Networks

A Bayesian network is a directed acyclic graphical (DAG) model that captures the joint distribution probability over a set of variables by a factorization in local conditional probabilities linking one random variable with its "parents." The fact that a Bayesian network naturally defines a directed graph makes this formalism highly suitable for learning directed GRNs and unsurprisingly, this mathematical model has already been used to predict GRN in the context of pure expression data analysis in the seminal paper (Friedman et al. 2000).

More formally, a Bayesian network denoted by $B = (\mathcal{G}, P_{\mathcal{G}})$ is defined by a DAG $\mathcal{G} = (V, A)$ with vertices representing random discrete variables $V = \{V_1, \ldots, V_m\}$, linked by a set of directed edges A, and a set of conditional probability distributions $P_{\mathcal{G}} = \{P_1, \ldots, P_m\}$. The variables involved in each conditional probability table P_i are defined by the DAG: $P_i = \mathbb{P}(V_i|Pa(V_i))$, where $Pa(V_i) = \{V_j \in V \mid (V_j, V_i) \in A\}$ is the set of parental nodes of V_i in $\mathcal{G}$.

The DAG of a Bayesian network B implicitly captures a set of conditional independencies between variables and represents a joint probability distribution on V defined as:

$$\mathbb{P}(V) = \prod_{i=1}^{m} \mathbb{P}(V_i|Pa(V_i)) \tag{2.5}$$

To model the available data, as in previous approaches, we used one variable E_i to represent the expression level of gene i and one variable M_i to represent the associated allelic state (for all $i \in \{1, \ldots, p\}$). All variables are discrete (see below for expression level discretization scheme) , allowing to capture nonlinear relationships between variables. If a given DAG structure G is assumed, maximum likelihood estimates of the parameters defining the conditional probability tables can be computed by simple counting. The GRN learning process then reduces to the problem of learning a DAG structure among these variables that maximizes $\mathbb{P}(\mathcal{G}|D) \propto \mathbb{P}(D|\mathcal{G})\mathbb{P}(\mathcal{G})$, where D represents the observed data.

Under specific assumptions, the marginal loglikelihood $\log(\mathbb{P}(D|\mathcal{G}))$ can be expressed as a decomposable scoring function. We used the BDeu score (Heckerman et al. 1995).

a statistical point of view, GRN learning is cursed by its high-dimensionality: the number of genes in a typical genome is much larger than the number of samples that can reasonably be produced ever. Existing works that decipher GRN structure based on genetical genomics data have been using Bayesian networks using genetic data as prior information (Zhu et al. 2007) or complex multivariate regression in a structural equation modeling framework with multiple testing, greedy search, and filtering steps (Liu et al. 2008). More recently, a meta-analysis of the output of different statistical methods targeted at learning in high dimension (based on penalized linear regression or penalized Bayesian network structure learning) has been shown to define the best performer (Vignes et al. 2011) on different datasets of simulated genetical genomics data, including up to 1,000 genes.

This chapter follows on this result in different directions: we first exploit the new genetical genomics simulated datasets that were produced by A. de la Fuente and colleagues, described in more detail in Chap. 1 of this book. These new datasets include a variety of different network topologies and include larger sets of genes (up to 5,000 genes for the largest ones), defining very challenging problems both in terms of dimensionality and in terms of computational learning cost. We also sophisticate and extend the set of statistical methods that are tested on these problems. The original methods included gene-by-gene Lasso regressions (Tibshirani 1996) and the Dantzig selector (Candès and Tao 2007) as well as a penalized Bayesian network learning algorithm. We have improved each of them by bootstrapping, in order to offer more reliable ranked list of edges representing regulations. Finally, we also integrated the random forest approach (Breiman 2001). This method has been used in Huynh-Thu et al. (2010) for GRN learning with expression data only. It is of specific interest for its ability to predict directed edges in all cases, compensating for the weaknesses of linear regression models, that only infer causal orientation from linear marker to genes relations, and Bayesian networks which may not allow to orient edges in all situations because of Markov equivalence (Koller and Friedman 2009).

In Sect. 2.2, we present the mathematical models and associated learning methods which have been used to analyze the data. In Sect. 2.3, we then present and discuss the results obtained by each of these methods and conclude.

2.2 Methods

In this section, we detail the statistical models and learning methods we used to tackle the datasets at hand, and how we adapted them to actually learn GNR with an associated edge-specific confidence score.

2.2.1 Data and Notations

The genetical genomics datasets were provided by Alberto de la Fuente and colleagues from CRS4 Bioinformatica (Pula, Italia). These datasets were simulated with the SysGenSim software (Pinna et al. 2011) and are available at the following http://sysgensim.sourceforge.net/datasets.html. Starting from a given network (a directed graph), the software simulates the network behavior using differential equations capturing expression, regulation, and molecule decay. The datasets and simulator are detailed in Chap. 1 of this book.

For each gene regulation network, a dataset is defined by a sample of n RILs which are measured for p bi-allelic markers and p gene expression levels. Every polymorphism is associated with a single gene and may influence either its direct expression (*cis* polymorphism occurring in the regulatory region of the gene) or its ability to regulate other target genes (*trans* polymorphism in the transcribed gene region itself, influencing its affinity with other gene regulatory complexes). A dataset is therefore defined by:

1. a $n \times p$ matrix $\mathbf{e}$ where e_{ij} gives the steady-state expression level of gene j for the RIL individual i (a real number). Each e_{ij} is an observation of the random variable E_j representing the expression level of gene j. $\mathbf{E}$ is the random matrix of all such variables. For a given gene $g \in \{1, \dots, p\}$ we denote by $\mathbf{E}^{-g}$, matrix $\mathbf{E}$ omitting its gth component.
2. a $n \times p$ matrix $\mathbf{m}$ where m_{ij} gives the allelic state of the polymorphism associated with gene j for RIL individual i (a $0/1$-data). Each m_{ij} is an observation of the random variable M_j representing the allelic state of the polymorphism associated to gene j. $\mathbf{M}$ is the random matrice of all such variables.

The dataset generation is controlled by the network size p, the RIL population size n, chromosome size, and gene expression heritability. The chromosome size is controlled by a mean genetic distance (either 1 or 5 cM) between adjacent marker positions, on all five chromosomes. The shorter the genetic distances between two markers, the stronger their genetic linkage. This leads to highly correlated allelic states between neighboring and even close markers. The latter parameter, heritability, is defined for each gene as the ratio of expression variance due to genetic factors over the expression variance when both genetic factors and biological/technical noise is accounted for. A broader distribution for biological noise implies lower gene heritability.

Our goal is to reconstruct the GRN that gave rise to the observed steady-state expression measures. Following the DREAM5 challenge, a prediction is defined by a directed consensus graph, each directed edge being associated with a "confidence score." In practice, all the statistical inference methods we used produce two confidence scores. One score derives from the estimated influence of allelic states on each expression levels and is denoted w^m_{kl} for the $k \to l$ edge. A similar score w^e_{kl} is obtained by estimating the influence of expression level E_k on expression level E_l.

In the following sections we introduce the most important components of the statistical learning methods we used: bootstrapping, penalized linear regressions, able to capture important interaction patterns, but parameter tuning or ad hoc pre- and post-processing may also have an important effect on the overall learning quality.

2.1 Introduction, Motivations

One of the central targets of *Systems Biology* is to decipher the complex behavior of a living cell in its environment. The effective behavior of the cell is probably defined through multiple layers of interacting entities including DNA, mRNA, non-coding RNA, proteins, and metabolites. In this chapter, we are interested in the so-called *genetical genomics* approach that combines the power of genetics, through the polymorphism, together with the measurement capabilities of gene expression to decipher a gene regulatory network (GRN), a simplified representation of the gene-level interactions occurring under given conditions. In such a network, vertices represent genes and directed edges represent the direct causal effect of genes over the expression of other genes through gene regulation (which can be activation or repression). Although proteins are often considered as the main vector of such regulations (through transcription factors for example), this simplified view of regulation could also accommodate other regulation effective transcribed molecules acting on transcription levels such as ncRNA genes. Moreover, protein levels are still quite difficult to measure.

By deciphering the set of gene regulations that are acting in a given context, one may be able to identify the most important, possibly indirect, players in the network that are capable of influencing a specific gene expression or phenotype of interest (Yvert et al. 2003), one may also link network structure to associated functional properties (Leclerc 2008; Marbach et al. 2009) and more generally understand the way gene interactions can control the overall cell behavior. A variety of mathematical formalisms, continuous or discrete (Boolean network Thomas 1973), defined over time (ordinary differential equations Bansal and di Bernardo 2007 or dynamic Bayesian networks Rau et al. 2010; Lèbre et al. 2010) or in stationary states (Friedman et al. 1999) have been proposed to represent the complex behavior of known gene regulation networks. In this chapter, we consider different statistical models of gene regulation that have been chosen for their ability to automatically infer gene regulations from expression data. As initially proposed by Jansen and Nap (2001), in order to integrate some causality in the inference process, we do not rely on time series of expression data but follow the so-coined *genetical genomics* (or *Systems Genetics*) path that exploits the possible influence of genetic polymorphisms on genic expression in a population (Aten et al. 2008). In a genetically controlled setting, defined for example by a population of recombinant inbred lines (RILs) producing randomly perturbed polymorphism combinations, the influence and genomic positions of polymorphisms that are observed in the population can be exploited to predict causal influences between gene expressions. The added value of having both genetic polymorphisms and perturbed phenotypic data, the expression level of genes, has already been demonstrated, in particular to infer causality (Zhu et al. 2007). From

Chapter 2
A Panel of Learning Methods for the Reconstruction of Gene Regulatory Networks in a Systems Genetics Context

David Allouche, Christine Cierco-Ayrolles, Simon de Givry, Gérald Guillermin, Brigitte Mangin, Thomas Schiex, Jimmy Vandel and Matthieu Vignes

Abstract In this chapter, we study different gene regulatory network learning methods based on penalized linear regressions (the Lasso regression and the Dantzig Selector), Bayesian networks, and random forests. We also replicated the learning scheme using bootstrapped sub-samples of the observations. The biological motivation relies on a tough nut to crack in Systems Biology: understanding the intertwined action of genome elements and gene activity to model gene regulatory features of an organism. We introduce the used methodologies, and then assess the methods on simulated "Systems Genetics" (or genetical genomics) datasets. Our results show that methods have very different performances depending on tested simulation settings: total number of genes in the considered network, sample size, gene expression heritability, and chromosome length. We observe that the proposed approaches are

---

D. Allouche · C. Cierco-Ayrolles · S. de Givry · G. Guillermin · B. Mangin · T. Schiex · J. Vandel · M. Vignes (✉)
UR875 MIA-T, INRA Toulouse, 31326 Castanet-Tolosan, France
e-mail: Matthieu.Vignes@toulouse.inra.fr

D. Allouche
e-mail: David.Allouche@toulouse.inra.fr

C. Cierco-Ayrolles
e-mail: Christine.Cierco-Ayrolles@toulouse.inra.fr

S. de Givry
e-mail: simon.degivry@toulouse.inra.fr

G. Guillermin
e-mail: Gerald.Guillermin@toulouse.inra.fr

B. Mangin
e-mail: Brigitte.Mangin@toulouse.inra.fr

T. Schiex
e-mail: Thomas.Schiex@toulouse.inra.fr

J. Vandel
e-mail: Jimmy.Vandel@toulouse.inra.fr

A. de la Fuente (ed.), *Gene Network Inference*,
DOI: 10.1007/978-3-642-45161-4_2,

Constant	$p = a$, where a is a real number.
Uniform distribution	$p = a + (b - a)\rho_u$, where a and b are the lower and upper limits of the uniform distribution $[a, b]$, and ρ_u is a random number sampled from the standard [0, 1] uniform distribution.
Normal distribution	$p = a + b\rho_n$, where a is the mean and b the standard deviation, and ρ_n is a random value drawn from the standard normal $N(0, 1)$ distribution. In the very unlikely (by choice of parameters a and b) case of $p < 0$, then $p = 0$ is forced to avoid negative parameters.
Gamma distribution	To guarantee a positive value for the parameters, a gamma distribution is the best choice. Parameters are randomly sampled from a Gamma distribution with density function:

$$\text{Gamma}(a, b) = \frac{1}{b^a \Gamma(a)} x^{a-1} e^{-x/b} \tag{1.2}$$

where $\Gamma(\cdot)$ is the gamma function, and a and b are the shape and scale parameters, respectively. The exponential distribution is a special case of the gamma distribution with $a = 1$.

References

Pinna A, Soranzo N, Hoeschele I, de la Fuente A (2011) Simulating systems genetics data with SysGenSIM. Bioinforma 27(17):2459–2462

Haldane JBS (1919) The combination of linkage values and the calculation of distance between the loci of linked factors. J Genet 8:299–309

Kosambi DD (1944) The estimation of map distances from recombination values. Ann Eugenics 12:172–175

c. Convert[10] the distances d to recombination rates r and compute the probability[11] p_k of no recombination between any adjacent markers k and $k+1$.
d. Generate the genotype vector for the entire chromosome $\mathbf{X}^h$ as:
   (i) Randomly set X_1 to 0 or 1 with equal probability.
   (ii) Sample u from a standard uniform distribution, set $X_k = X_{k-1}$ if $u < p_k$ and otherwise set $X_k = 1 - X_{k-1}$.

3. Combine all $\mathbf{X}^h$ into one single genotype vector $\mathbf{X}$.
4. The allelic effects (see Eq. 1.1 in Sect. 1.3.3) of the *cis* (c) and *trans* (t) variants are generated as follows:

   a. For variant i (variant is synonymous with gene here as each gene is only allowed to have a single functional variant in *cis* or *trans*), set $Z_i = 1$. Sample u from a standard uniform distribution and if $u > 0.5$, sample Z_i from the uniform distribution $[Z^l, Z^u]$.
   b. Randomly select $n \cdot p_{cis}$ genes i to have a *cis* variant and set $Z_i^c = Z_i$.
   c. For the remaining genes j having a *trans* variant, set $Z_j^t = Z_j$.

5. A pre-specified number (proportion) of genotyping errors are added by changing randomly selected entries of the genotype vector $\mathbf{X}$ from 0 to 1 or vice versa.

1.3.3 Simulation of the Gene Expression Data

For all the individuals, SysGenSIM computes the solution of a system composed of n differential equations, one for each gene i:

$$\frac{\mathrm{d}G_i}{\mathrm{d}t} = V_i Z_i^c \theta_i^{\mathrm{syn}} \prod_{j \in \mathscr{R}_i} \left[1 + A_{ji} \frac{G_j^{h_{ji}}}{G_j^{h_{ji}} + \left(K_{ji}/Z_j^t\right)^{h_{ji}}} \right] - \lambda_i \theta_i^{\mathrm{deg}} G_i \qquad (1.1)$$

where $\mathscr{R}_i$ is a set containing the indices of all regulators (both activators and inhibitors) j of gene i; G_i is the gene expression value of gene i, V_i is its basal transcription rate, and λ_i its degradation rate constant. K_{ji} is the interaction strength of G_j on G_i, h_{ji} is the Hill cooperativity coefficient, and A_{ji} is an element of the adjacency matrix $\mathbf{A}$ encoding the signed network structure. Finally, the parameters θ_i^{syn} and θ_i^{deg} represent the biological variances in the synthesis and degradation processes of gene i, while Z_i^c (cis-effect) and Z_i^t (trans-effect) incorporate the effects of DNA polymorphisms in the model.

Except for the case of a constant, the distributions from which the model parameters p are sampled are defined by two parameters a and b. The possible distributions are listed below:

---

[10] According to Haldane: $r = 0.5(1 - e^{-0.02d})$.

[11] For recombinant inbred lines generated by selfing inbred line cross: $p = 1/(1 + 2r)$.

Table 1.3 Settings applied to each network

Configuration	Marker distance	Biological variance	Heritability	Population size
1	N(5, 1)	N(1, 0.1)	≃0.8	300
2	N(5, 1)	N(1, 0.1)	≃0.8	900
3	N(5, 1)	N(1, 0.25)	≃0.4	300
4	N(5, 1)	N(1, 0.25)	≃0.4	900
5	N(1, 0.1)	N(1, 0.1)	≃0.8	300
6	N(1, 0.1)	N(1, 0.1)	≃0.8	900
7	N(1, 0.1)	N(1, 0.25)	≃0.4	300
8	N(1, 0.1)	N(1, 0.25)	≃0.4	900

1.3 Algorithms in SysGenSIM

This section is dedicated to the detailed description of the most relevant algorithms underlying the whole simulation process. In particular, the procedure to build the EIPO *in silico* networks is illustrated in Sect. 1.3.1, the simulation of the genotypes is described in Sect. 1.3.2, and the equation modeling the evolution of the gene activity is explained in Sect. 1.3.3.

1.3.1 Generation of EIPO Networks

The algorithm generates a network with exponential in-degree and power law out-degree node distributions,[6] requiring the user to only specify the size n and the desired average degree K_d. Subsequently, the inverse scale parameter of the exponential distribution[7] is set to $\lambda = 1/K_d$, while the exponent γ in the power law distribution[8] is found after a quick iterative search. The algorithm works as follows:

1. Set $\lambda = 1/K_d$ defining the exponential distribution.
2. Find γ and hence the power law distribution according to the desired average degree K_d.
3. Calculate the discrete probabilities $\mathbf{P}_{\text{in}}$ and $\mathbf{P}_{\text{out}}$ from the two distributions, respectively, for each possible degree K, i.e. from $K = 0$ to $K = n - 1$ (a node can maximally be linked to all the remaining $n - 1$ nodes).
4. Initialize the adjacency matrix $\mathbf{A}$ to zero.
5. Sample the in- and out-degree arrays $\mathbf{K}_{\text{in}}$ and $\mathbf{K}_{\text{out}}$ from the respective distributions, i.e. according to the probabilities $\mathbf{P}_{\text{in}}$ and $\mathbf{P}_{\text{out}}$. The i-th entries of the arrays

[6] In-degree and out-degree refer to the number of ingoing and outgoing edges of a node in a graph, respectively.

[7] The probability density function of an exponential distribution is $f(x; \lambda) = \lambda e^{-\lambda x}$ for $x \geq 0$.

[8] The power law distribution is described by the probability density function $f(x; \gamma) = x^{-\gamma}$, for $x \geq 0$.

$\mathbf{K}_{\text{in}}$ and $\mathbf{K}_{\text{out}}$ represent, respectively, the number of ingoing and outgoing edges for node i. The sampling is performed until the following conditions are met:

- $K_{\text{in}}^i + K_{\text{out}}^i > 0, \forall i$, i.e. all nodes are connected to the rest of the network through at least one (ingoing or outgoing) edge.
- $\sum_i K_{\text{in}}^i \geq \sum_i K_{\text{out}}^i \simeq n \cdot K_d$, to assure[9] that all the outgoing edges from $\mathbf{K}_{\text{out}}$ can reach a node in $\mathbf{K}_{\text{in}}$.

6. Sort the nodes with $K_{\text{out}}^i > 0$ (i.e. with at least one outgoing edge) by descending out-degree in list $\mathcal{N}_{\text{out}}$.
7. Then, for each node i in the ordered list $\mathcal{N}_{\text{out}}$:

   a. Place the nodes j with positive in-degree K_{in}^j in set $\mathcal{N}_{\text{in}}$.
   b. Remove, if included, node i from $\mathcal{N}_{\text{in}}$.
   c. If $|\mathcal{N}_{\text{in}}| < K_{\text{out}}^i$ then go back to step 5, else connect node i with K_{out}^i randomly selected nodes from $\mathcal{N}_{\text{in}}$.
   d. Decrease by 1 the in-degree of the nodes that have been just connected to node i.
   e. Update the adjacency matrix $\mathbf{A}$ with the new edges.

After the procedure has been performed for all nodes in $\mathcal{N}_{\text{out}}$, the network is complete and exhibits exponential in-degree and power law out-degree distributions with average degree $K \simeq K_d$.

1.3.2 Simulation of the Genotypes

SysGenSIM simulates genotype data according to a user-defined or to a randomly generated genetic map based on the number of chromosomes in the genome and the number of genetic markers per chromosome with constant or normally distributed pair-wise distance among DNA variant locations. Map distances are converted to recombination rates for the generation of genotypes at ordered linked loci through the Haldane (1919) or Kosambi (1944) mapping functions. Markers can be either placed in perfect linkage with each functional polymorphism, or a marker map can be generated and then the functional variants randomly placed throughout the genome.

The genotype data for the benchmark datasets have been generated by the following algorithm:

1. Sample the number of markers for the n_{cr} chromosomes according to the size of the network n and the selected distribution of markers per chromosome.
2. Then, for each chromosome h in the genome:

   a. Generate the marker distances d by sampling the values according to the selected distribution ($N(1, 0.1), N(5, 1)$).
   b. Map the functional polymorphisms at markers.

---

[9] The condition is requested in the continuation of the algorithm.

Alberto de la Fuente
Editor

Gene Network Inference

Verification of Methods for Systems Genetics Data

Springer

Editor
Alberto de la Fuente
Leibniz-Institute for Farm Animal Biology
Dummerstorf
Germany

ISBN 978-3-642-45160-7 ISBN 978-3-642-45161-4 (eBook)
DOI 10.1007/978-3-642-45161-4
Springer Heidelberg New York Dordrecht London

Library of Congress Control Number: 2013957428

Printed on acid-free paper

Springer is part of Springer Science+Business Media (www.springer.com)

Complex issues like SG data analysis need to be addressed as a community rather than by individual groups. This book provides the result of an international community effort dedicated to evaluating algorithms for the inference of Gene Regulatory Networks from SG data. Before the application of learning algorithms to SG data, they need to be thoroughly evaluated on benchmark datasets. Thorough and fair assessment of algorithms is of utmost importance to identify their merits and pitfalls, before eventually extracting biological insights from SG data. These studies will only be successful if powerful data analysis methods are available and applied correctly. To conclude, I would like to acknowledge the European COST Action STATSEQ for financing a meeting among the authors of the book chapters in order to present and discuss all the methods presented in this book.

The benchmarks studied in this book, the "StatSeq benchmark dataset," are available to the community at: http://sysgensim.sourceforge.net/datasets.html.

Germany Alberto de la Fuente

References

Jansen RC (2003) Studying complex biological systems using multifactorial perturbation. Nat Rev Gen 4:145–151

Liu B, Hoeschele I et al (2009) Inferring gene regulatory networks from genetical genomics data. In: Das S, Caragea D, Welch SM, Hsu WH (eds) Handbook of research on computational methodologies in gene regulatory networks. IGI Global, New York, p 79–107

Marbach D, Costello JC et al (2012) Wisdom of crowds for robust gene network inference. Nat Methods 9(8):96–804

Meyer P, Alexopoulos LG et al (2011) Verification of systems biology research in the age of collaborative competition. Nat Biotechnol 29(9):811–815

Pinna A, Soranzo N et al (2011) Simulating systems genetics data with SysGenSIM. Bioinformatics 27(17):2459–2462

Rockman MV (2008) Reverse engineering the genotype-phenotype map with natural genetic variation. Nature 456(7223):738–744

Stolovitzky G, Monroe D et al (2007) Dialogue on reverse-engineering assessment and methods: the DREAM of high-throughput pathway inference. Ann NY Acad Sci 1115:1–22

Wang Z, Gerstein M et al (2009) RNA-Seq: a revolutionary tool for transcriptomics. Nat Rev Genet 10(1):57–63

Chapter 1
Simulation of the Benchmark Datasets

Andrea Pinna, Nicola Soranzo, Alberto de la Fuente and Ina Hoeschele

Abstract In this chapter, the *in silico* systems genetics dataset, used as a benchmark in the rest of the book, is described in detail, in particular regarding its simulation by SysGenSIM. Morever, the algorithms underlying the generation of the gene expression data and the genotype values are fully illustrated.

1.1 Introduction

The presented benchmark dataset is meant to be used for training and evaluating algorithms and techniques for the inference of networks from systems genetics data. The goal is to find which methodologies exhibit the best overall network inference performance, and to analyze their performances under particular conditions (i.e. population size, large or small marker distances, high or low heritability, and network size).

---

[1] SysGenSIM 1.0.2, version released on May 8th, 2012. More information is available in the online manual at http://sysgensim.sourceforge.net.

A. Pinna · N. Soranzo
CRS4 Bioinformatica, 09010, Pula, Italy
e-mail: andrea.pinna@crs4.it

N. Soranzo
e-mail: soranzo@crs4.it

I. Hoeschele
Virginia Polytechnic Institute and State University,
VA 24061, Blacksburg, USA
e-mail: inah@vbi.vt.edu

A. de la Fuente (✉)
Institute of Genetics and Biometry, Leibniz-Institute for Farm Animal Biology,
18196, Dummerstorf, Germany
e-mail: alf@fbn-dummerstorf.de

A. de la Fuente (ed.), *Gene Network Inference*,
DOI: 10.1007/978-3-642-45161-4_1, © Springer-Verlag Berlin Heidelberg 2013

Section 1.2 describes how the dataset has been generated by means of SysGenSIM,[1] a MATLAB toolbox for simulating systems genetics experiments in model organisms Pinna et al. (2011). Detailed information is provided about the topology of the gene networks and the settings of the simulator used to produce the genotypes and the gene expression data. In Sect. 1.3, the algorithms employed to obtain the networks and genetic data are thoroughly explained.

1.2 Description of the Systems Genetics Dataset

The Systems Genetics benchmark is a collection of 72 *in silico* datasets generated from nine artificial gene networks of different size. In the following, details on the *in silico* networks (Sect. 1.2.1) and on the configurations of SysGenSIM employed to produce the data (Sect. 1.2.2) are provided.

1.2.1 In Silico Networks

The systems genetics experiments have been simulated using nine different artificial gene networks. These networks have been generated using SysGenSIM with parameters chosen to produce the following topological properties:

- three networks for each of the sizes $n = \{100, 1000, 5000\}$, in the following referred to with the self-explanatory labels 100-{1, 2, 3}, 1000-{1, 2, 3}, and 5000-{1, 2, 3};
- exponential in-degree and power law out-degree (EIPO) distributions for the nodes;
- average node degree[2] $K \simeq 6$;
- size of the largest strongly connected component (sub-network) equal to at least 20 % of the network nodes ($n = 100$), 15 % ($n = 1000$), and 10 % ($n = 5000$).

Some other topological characteristics of the nine networks are summarized in Table 1.1, where for each of the networks the number of edges, the size of the largest strongly connected component (LSCC), the number of nodes in the in- and in the out-components,[3] and the number of nodes amongst tendrils[4] and tubes[5] are shown. Network 1000-2 has one isolated node. Figure 1.1 shows a representation of such network topology.

---

[2] The average number of both ingoing and outgoing edges for a node: $K = K_{\text{in}} + K_{\text{out}}$.

[3] Respectively, all the nodes from which the LSCC is reachable and that are not reachable from the LSCC, and all the nodes reachable from the LSCC but from which the LSCC cannot be reached.

[4] Nodes from which the LSCC cannot be reached, and that cannot be reached from the LSCC itself.

[5] Nodes connecting the in- to the out-component, and not belonging to the LSCC.

Table 1.1 Topological characteristics of the *in silico* networks

Size	Network label	Edges	LSCC	In-	Out-	Tendrils	Tubes
100	100-1	285	21	25	18	30	6
100	100-2	304	26	7	63	4	0
100	100-3	296	34	9	57	0	0
1000	1000-1	3149	165	71	660	96	8
1000	1000-2	2881	162	58	648	108	23
1000	1000-3	3038	150	106	563	158	23
5000	5000-1	14678	528	224	3498	657	93
5000	5000-2	15672	526	233	3580	595	66
5000	5000-3	15270	598	231	3660	480	31

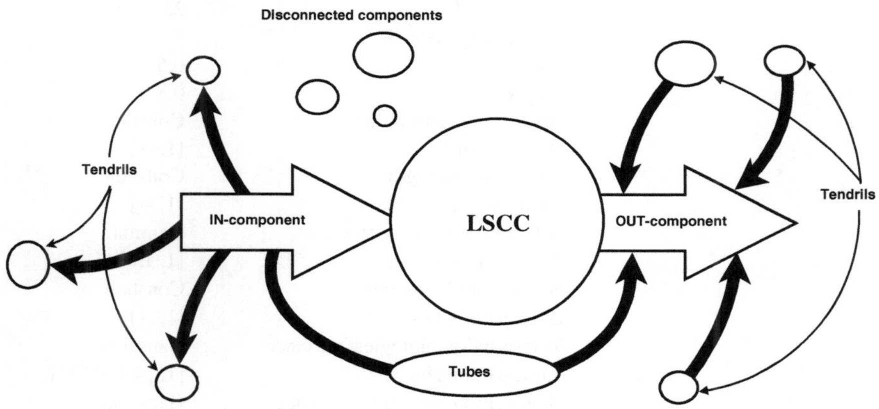

Fig. 1.1 Model representing the topology of the artificial networks

1.2.2 Simulation of Datasets

In order to provide a wide range of scenarios, datasets were simulated by combining the nine topologies with eight different parameter settings for a total of 72 datasets. The eight parameter settings resulted from a combination of two average marker distances $d = \{1, 5 \text{ cM}\}$, two median heritability values H (high $\simeq 0.8$ and low $\simeq 0.4$), and two population sizes $m = \{300, 900\}$. Simulations have been run with SysGenSIM's optional parameter settings set as described in Table 1.2.

By keeping most of the parameters fixed, each dataset has been simulated according to the setting configurations summarized in Table 1.3, i.e. only the marker distance, the biological variance, and the population size have been manipulated.

For each of the 72 datasets, the following four components were made available, the first two for data analysis (network inference) and the other two for algorithm evaluation purposes:

Table 1.2 Values of SysGenSIM's optional parameter settings used to generate the datasets

Settings	Selected value
Network topology	EIPO
Network size	{100, 1000, 5000}
Edge sign assignment	Node-wise
Edge sign probability	0.5
Average node degree	6
Marker positions	Generate
Gene positions	At markers
Mapping function	Haldane
RIL type	Selfing
Number of chromosomes	{5, 25, 25}
Markers per chromosome	N({20, 40, 200}, 2)
Marker distances	{N(1, 0.2), N(5, 1)}
Cis-effect %	25
Genotyping error %	5
Z lower	0.5
Z upper	0.8
Basal transcription rate	Constant
B.t.r. parameters	[1, −]
Interaction strength	Constant
I.s. parameters	[1, −]
Hill cooperativity coefficient	Gamma
H.c.c. parameters	[1, 1.67]
Basal degradation rate	Constant
B.d.r. parameters	[1, −]
Transcription biological variance	Gaussian
T.b.v. parameters	[1, {0.1, 0.25}]
Degradation biological variance	Gaussian
D.b.v. parameters	[1, {0.1, 0.25}]
Expression measurement noise	Gaussian
E.m.n. parameters	[1, 0.1]
Number of phenotype nodes	0
Population size	{300, 900}
Number of experiments	1

Gene expression matrix — A $n \times m$ matrix containing gene expression measurements. Entry (i, j) is the simulated steady-state expression value of gene i in individual j.

Genotype matrix — A $n \times m$ matrix of genotype values $\{0, 1\}$. Entry (i, j) is the genotype value of gene i in individual j.

Heritability — The median value of the heritability.

Edge list — A signed list of edges encoding the direct interactions between the nodes of the network.

Contents

1 **Simulation of the Benchmark Datasets** 1
Andrea Pinna, Nicola Soranzo, Alberto de la Fuente
and Ina Hoeschele

2 **A Panel of Learning Methods for the Reconstruction of Gene Regulatory Networks in a Systems Genetics Context** 9
David Allouche, Christine Cierco-Ayrolles, Simon de Givry,
Gérald Guillermin, Brigitte Mangin, Thomas Schiex,
Jimmy Vandel and Matthieu Vignes

3 **Benchmarking a Simple Yet Effective Approach for Inferring Gene Regulatory Networks from Systems Genetics Data** 33
Sandra Heise, Robert J. Flassig and Steffen Klamt

4 **Differential Equation Based Reverse-Engineering Algorithms: Pros and Cons** 49
Gennaro Gambardella, Roberto Pagliarini, Francesco Gregoretti,
Gennaro Oliva and Diego di Bernardo

5 **Gene Regulatory Network Inference from Systems Genetics Data Using Tree-Based Methods** 63
Vân Anh Huynh-Thu, Louis Wehenkel and Pierre Geurts

6 **Extending Partially Known Networks** 87
Pegah Tavakkolkhah and Robert Küffner

7 **Integration of Genetic Variation as External Perturbation to Reverse Engineer Regulatory Networks from Gene Expression Data** 107
Francesco Sambo, Tiziana Sanavia and Barbara Di Camillo

8 **Using Simulated Data to Evaluate Bayesian Network Approach for Integrating Diverse Data** 119
Luan Lin and Jun Zhu

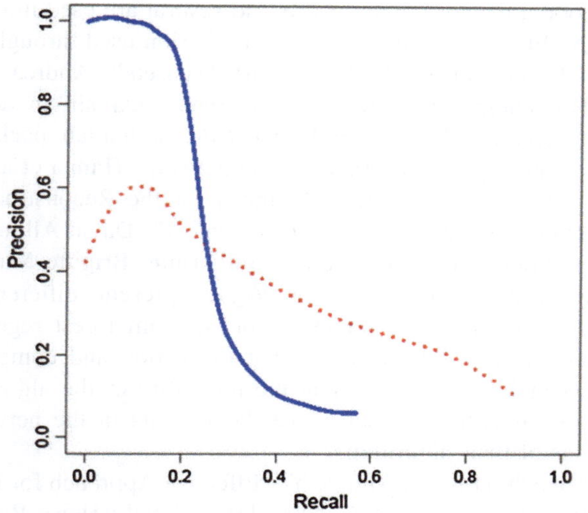

Fig. 2 Method I (*solid line*) predicts the true edges well in the top of the list (high precision), but then accumulates many false positives, sharply dropping precision, at increasing recall. Method II (*dotted line*) has lower precision in the top of the list, but the precision remains relatively high at increasing recall

existence of that edge. The more the true edges in the network occur in the top of the predicted ranked list, the better the inference algorithm is considered to be.

The two important quantities can be expressed as

$$\text{Pecision} = \frac{TP}{TP+FP}$$
$$\text{Recall} = \frac{TP}{TP+FN}$$

where *TP* refers to True Positives, edges that are correctly predicted, *FP* refers to False Positives, edges that are predicted but actually don't occur in the true network, and *FN* refers to False Negatives, the edges that do occur in the true network, but are not predicted. The Precision provides the fraction of correct edges among the predicted edges and the Recall provides the fraction of true edges that has been discovered.

To obtain a "Precision versus Recall curve," the Precision and Recall are evaluated from the top of the sorted list of predicted edges, each time evaluating for a network with one edge more included. This way a curve is obtained with as many points as there are edges in the prediction list. The larger the area under the Precision versus Recall, the better the algorithm used to make the predictions is. The maximum value for the AUPR is one (perfect prediction: all true edges are ranked at the top of the list) and the minimum value is 0 (worst possible prediction: all true edges are ranked at the bottom of the list). Figure 2 shows the precision versus recall curve of two distinct methods. There are many alternative ways to evaluate how good a network prediction is and several are used in the chapters in this book, but AUPR is the central evaluation criterion used throughout the book.

In Chap. 1 "Simulation of the Benchmark Datasets", Andrea Pinna, Nicola Soranzo, Ina Hoeschele, and Alberto de la Fuente explain in detail how the benchmark was created. They used SysGenSIM, a matlab package specially designed to simulate SG data with large gene networks (Pinna et al. 2011).

In Chap. 2 "A Panel of Learning Methods for the Reconstruction of Gene Regulatory Networks in a Systems Genetics Context", David Allouche, Christine Cierco-Ayrolles, Simon de Givry, Gerald Guillermin, Brigitte Mangin, Thomas Schiex, Jimmy Vandel, and Matthieu Vignes present different approaches (penalized regression, Bayesian networks, and random forest regression) to the benchmark datasets and perform an elegant evaluation and comparison among these methods. Notably, they show that the reliability of the algorithms heavily depended on several criteria to determine the weights in the network from the initial predictions of their algorithms.

In Chap. 3 "Benchmarking a Simple Yet Effective Approach for Inferring Gene Regulatory Networks from Systems Genetics Data", Sandra Heise, Robert J. Flassig, and Steffen Klamt demonstrate their novel network inference approach on the benchmark. Based on the marker-expression associations they first build an initial "perturbation graph," which is subsequently sparsified using transitive reduction.

In Chap. 4 "Differential Equation-Based Reverse-Engineering Algorithms: Pros and Cons", Gennaro Gambardella, Roberto Pagliarini, Francesco Gregoretti, Gennaro Oliva, and Diego di Bernardo apply a linear differential equation model to the benchmark data. The approach was originally conceived to analyze experimental perturbation data, but could be applied to SG data as well.

In Chap. 5 "Gene Regulatory Network Inference from Systems Genetics Data Using Tree-Based Methods", Vân Anh Huynh-Thu, Louis Wehenkel, and Pierre Geurts modified their network inference algorithm GENIE3, based on random forest regression, to be able to perform integrative analysis of genotypes and gene expression.

In Chap. 6 "Extending Partially Known Networks", Pegah Tavakkolkhah and Robert Küffner present a method to infer gene networks which relies on prior knowledge of the network. Assuming knowledge of 50 % of the edges, they applied a supervised inference framework to the provided expression datasets and genotype information.

Chapter 7 "Integration of Genetic Variation as External Perturbation to Reverse Engineer Regulatory Networks from Gene Expression Data" by Francesco Sambo, Tiziana Sanavia, and Barbara Di Camillo describe a method based on genotype-gene expression correlations and a subsequent block-prioritization approach to disentangle true causal effects from spurious correlations.

Chapter 8 "Using Simulated Data to Evaluate Bayesian Network Approach for Integrating Diverse Data", Luan Lin and Jun Zhu present their algorithm Reconstructing Integrative Molecular Bayesian Networks (RIMBANet), a tool based on Bayesian framework which permits both linear and nonlinear regulation identification and which facilitates the incorporation of prior knowledge.

Foreword

Phenotypic variation, including that which underlies health and disease in humans, often results from multiple interactions among numerous genetic and environmental factors. Systems Genetics (SG) seeks to understand this complexity by integrating the questions and methods of systems biology with those of genetics to solve the fundamental problem of interrelating genotype and phenotype in complex traits and diseases (Jansen 2003; Rockman 2008; Liu et al. 2009). SG has been made possible by the advance of experimental techniques that allow measuring biological compounds all along the molecular hierarchy from DNA, RNA, proteins, metabolites, fatty acids to whole organism phenotypes.

Systems Genetics data consist of genotyping data and other datasets that potentially reflect the effect of a perturbation of the system caused by these naturally diverse genotypes: phenotypes of interest (e.g., disease, biomass yield), "molecular phenotypes" such as omics datasets: gene expression levels, gene methylations, proteins, and metabolite levels. Regular genetic studies (with genotyping and phenotyping data alone) permit the identification of genetic loci which affect a given phenotype. The availability of measurements of tens of thousands of molecular phenotypes enables algorithms to elucidate the regulatory networks underlying the complex genotype–phenotype relationships (Jansen 2003; Rockman 2008; Liu et al. 2009). Figure 1 illustrates the essential difference between statistical analysis of regular genetic studies and the approach of SG.

A large number of network inference methods for SG data have been proposed (Liu et al. 2009), and more algorithms are expected to appear now that SG data availability increases due to growing use of Next Generation Sequencing (Wang et al. 2009). Therefore, thorough verification and fair assessment of algorithms is of high importance to learn which algorithms are most useful for extracting biological insights from SG data. This issue needs to be addressed as a community effort (Meyer et al. 2011). In this book, recent methods for SG data analysis are described and applied to a suite of simulated SG benchmark datasets. Each of the chapter authors received the same datasets to evaluate the performance of their method.

The knowledge gained from such benchmarking study ultimately allows for a confident use of these algorithms in SG studies of, e.g., complex human diseases or food crop improvement. Several international initiatives emerged with the goal to systematically assess proposed algorithms in computational biology. Well-known examples include the Dialogue for Reverse Engineering Assessments and Methods

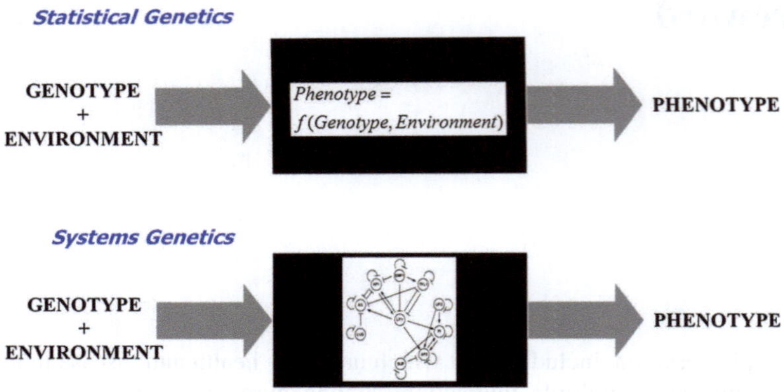

Fig. 1 The difference between statistical analysis of regular genetic studies and the approach of SG. While the former results in a "black box" model that relates phenotypes to genotype and environment by an abstract mathematical function, the latter aims to identify the gene networks establishing that relationship by making use of large-scale measurements of molecular phenotypes

(DREAM) project network reverse-engineering challenges (Stolovitzky et al. 2007; Marbach et al. 2012) (http://www.the-dream-project.org/), the Critical Assessment of protein Structure Prediction (CASP, http://predictioncenter.org/) tools and the Genetic Analysis Workshops (GAWs, http://www.gaworkshop.org/). Such initiatives have created a sort of "collaborative-competition" from which thorough insights for data analysis and novel biological discoveries have been obtained (Meyer et al. 2011). In this book, we take such a collaborative approach: each chapter describes methods and their evaluation on a common benchmark dataset. Benchmarking on real biological data is challenging as true regulatory networks are largely unknown. The availability of realistically simulated datasets, which are generated under a set of assumptions most relevant to real data, is of utmost importance for the verification of algorithms for data analysis. Only for these data we are certain about the true complex system underlying the data (Pinna et al. 2011).

The benchmark dataset used in this book is simulated using the software SysGenSIM (Pinna et al. 2011). Several parameter settings have been selected to test how inference is affected by important aspects as the number of observations, the heritability, genetic linkage, and network size. A total of 72 datasets were generated. In addition to the simulated gene expression and genotyping data, also the true network structures underlying the data were provided to give each of the chapter authors the ability to study the effects of different decisions in their inference approach (parameter values, scaling methods, etc.) and to get the maximum out of their algorithms.

There is a variety of approaches to evaluate how good a network prediction is. The authors were asked to use the "Area Under the Precision versus Recall curve" (AUPR) to evaluate their approaches. In this type of evaluation, instead of the prediction of an explicit network, the predictions are in the form of a ranked list of edges. The higher the rank of the edge, the more confidence is placed in the correlation between regressors excluded from the model and the regression residual. Interpreting t of Eq. (2.2) is also possible by considering the specific value $t_0 = || \hat{\theta}^{rss} ||_{\ell_1}$. Then, setting t to $t_0/2$ roughly shrinks active coefficients in the regression by 50 % (Hastie et al. 2009).

Equation (2.1) was solved using the Least Angle Regression (LAR) algorithm implemented in the R `glmnet` package (Friedman et al. 2010). The choice of the penalty λ is presented later.

2.2.3.2 The Dantzig Selector

The Dantzig selector (Candès and Tao 2007) is a related penalized linear regression method based on ℓ_1 norm penalization of the parameters subject to a constraint bound on the maximum absolute correlation between the residuals and regressors (we use ${}^{\top}X$ to denote the transpose of X):

$$\hat{\theta}^{\text{dantzig}} = \arg\min_{\theta} || \theta ||_{\ell_1}, || {}^{\top}X(Y - X\theta) ||_{\ell_\infty} \leq \delta, \tag{2.3}$$

where δ is the actual bound on the correlation between the residual and each regressor. With no bound, the Dantzig selector sets all coefficients to zero which minimizes the ℓ_1 norm of the parameters. When the bound tends to 0, the Dantzig selector imposes a null correlation between the residual and the regressors. This condition is satisfied by the RSS estimate, as it is equivalent to enforcing a null derivative of the RSS (Hastie et al. 2009). Equation (2.3) can be written in its dual form, as an analog of Eq. (2.2) for the Lasso:

$$\hat{\theta}^{\text{dantzig}} = \arg\min_{\theta} || {}^{\top}X(Y - X\theta) ||_{\ell_\infty}, || \theta ||_{\ell_1} \leq t \tag{2.4}$$

In Vignes et al. (2011), we used the reduction of the Dantzig selector to linear programming and an open source linear programming solver (glpk) for resolution. Because of the increased computational burden generated by bootstrapping, we decided to instead use a dedicated homotopy Dantzig algorithm (Asif and Romberg 2010) and its companion Matlab package, which was run using Octave. The choice of the value for the parameter δ is difficult and is described in the following.

2.2.3.3 Confidence Scores with Penalized Linear Regressions and Bootstrap

We want to provide confidence scores on the prediction of every oriented edge $j \rightarrow g$ capturing the causal influence of gene j on gene g. For a fixed value of the penalization and for a given bootstrap sample, two distinct cases can be identified. If α_{gj} estimation is nonzero, marker j is assumed to have a direct effect on the expression of gene g. The converse is impossible since expression levels cannot affect polymorphism. The interpretation of a nonzero β_{gj} (or β_{jg}) is slightly different: this indicates that a relationship exists between the expressions of genes j and g, but the causal orientation is unknown: either j influences g or g influences j.

Choosing the 'right' level of penalization in the Lasso regression or in the Dantzig selector is a difficult model selection problem. To avoid this problem, we follow the idea of Vignes et al. (2011), described here for the Lasso regression. Instead of choosing a fixed value for the penalty term λ, we explore an evenly spaced grid of possible penalty values from a starting value 0 (no penalization) to a maximum value $\lambda_{\max}$: the infimum of the set of all λ that prevents a single regressor to be included in any of the regressions. A total of $q = 10$ different penalty values we used from $\frac{\lambda_{\max}}{q}$ (low penalty level) to $\lambda_{\max}$ (maximal penalty level). A similar mechanism, with the same number q of penalty values, is used for the Dantzig selector. The fraction of times, over all penalizations, that a regressor is introduced with a nonzero parameter estimate was then used as a confidence score.

In our case, bootstrapping offers a second dimension that can be exploited to evaluate a confidence score. Besides the first dimension defined by the grid of q evenly spaced values of penalizations, a second dimension is available through the set of N_{boot} different bootstrap samples. We denote by $\#(\alpha_{gj})$ (resp. $\#(\beta_{gj})$) the total number of regressions along these two dimensions where $\alpha_{gj} \neq 0$ (resp. $\beta_{gj} \neq 0$). The marker-based confidence score w^m_{kl} of the oriented edge $k \rightarrow l$ is then defined as the frequency:

$$w^m_{kl} = \frac{\#(\alpha_{lk})}{q\, N_{\text{boot}}}$$

The computation of the expression-based confidence score w^e_{kl} of the oriented edge $k \rightarrow l$ has to take into account the fact that it can be derived from any (or both) of β_{kl} and β_{lk} to be nonzero and also that this information is uncertain on the two possible edge orientations. This leads to:

$$w^e_{kl} = \frac{\#(\beta_{kl}) + \#(\beta_{lk})}{4\, q\, N_{\text{boot}}}$$

Bach (2008) studied the asymptotic ($n \rightarrow \infty$) properties of the Lasso variable selection for some penalization decays. In specific settings, the Lasso tends to select correct variables with probability 1 and irrelevant variables with a probability which is strictly between 0 and 1. Hence Bach (2008) proposed to use bootstrapping to assess the probability of selecting a variable at a given penalty level, keeping only those variables that are always selected.

Our strategy is quite different from the strategy proposed in Bach (2008), and it may ultimately include false positive regressors in the model (especially for low confidence scores). But the properties of the analyzed data—including their high dimensionality—and the predicted object, with edge confidence scores, are different from those considered in Bach (2008).

We tested the Lasso method with values of N_{boot} equal to 100 and 200. Given the very limited impact of this choice on the results, we ultimately decided to use the computationally favorable solution of $N_{\text{boot}} = 100$.

2.2.5.1 GRN Structure Learning with Bootstrapped Greedy Search

Learning Bayesian networks is an NP-hard problem with a super-exponential search space of potential DAG structures (Chickering et al. 2004). Even a greedy search heuristic method can be very time-consuming when the number of variables p is large. In order to get reasonable computation times, we selected for each gene a set of candidate parents if the local BDeu score increases when each candidate parent is added separately in comparison to the empty graph. Furthermore, we kept only one marker having the best BDeu score increase in a sliding window of 10 markers along the chromosomes. We also limited the maximum number of parents per gene to 5. In addition, we took into account biological knowledge of the observed data to reduce the search space.

First, genetic linkage between close markers induces strong correlations between their allelic states. Learning the corresponding edges between marker variables M_i is useless for reconstructing the GRN and just makes the structure search more complex. We therefore forbade such edges. We also forbade edges from genes to markers, which have no biological meaning.

Furthermore, a preliminary analysis of variance was used to predict *cis*-regulatory markers: detected positive markers (Bonferroni corrected p-value < 0.1) were those giving the most significant signal in a seven marker-width window, centered on the gene, to avoid false marker influence due to genetic linkage. We used this *cis*-effect information to constrain the structure: since each *cis*-marker M_i had an effect on its associated gene activity E_i only, we constrained our model to use an $M_i \rightarrow E_i$ edge. In the opposite case, when the marker M_i was detected as not being a *cis*-regulatory marker, we only forbade the $M_i \not\rightarrow E_i$ edge.

The structure and parameters of the underlying graph can then be estimated using the BDeu score-based structure learning algorithm described in Vandel et al. (2012), using the same adaptive discretization policy into 2–4 states as in Vignes et al. (2011). We selected the DAG with the best BDeu score among three restarts of the Stochastic Greedy Search algorithm which exploits extended local move operators (SGS^3, Vandel et al. 2012). We used BDeu *equivalent sample size* parameter $\alpha = 1$. Each learnt DAG structure produces a set of oriented edges which therefore translate directly into the predicted GRN. For each edge $M_i \rightarrow E_j$ or $E_i \rightarrow E_j$ in the learnt structure, we predict the existence of a directed edge $i \rightarrow j$ in the GRN. Notice that thanks to this mapping procedure, an initial acyclic directed graph may ultimately lead to the prediction of a cyclic GRN.

This procedure was improved by bootstrapping, allowing to produce a set of directed edges with confidence scores set to the frequency of the corresponding edge over all bootstrap samples ($N_{\text{boot}} = 100$). Again, this may lead to cycles in the predicted GRN. Because of the important computational burden generated by bootstrapping, the Bayesian network approach could, however, not be applied to the largest datasets with $p = 5{,}000$ genes.

2.2.6 Postprocessing

Each of the methods produces two weighted oriented graphs, one that connects marker variables M_j to expression variables E_g (where edge $j \rightarrow g$ is weighted by w^m_{jg}) and another connecting expression variables E_j to other expression variables E_g (where edge $j \rightarrow g$ is weighted by w^e_{jg}). Ultimately,a single ranked list of edges is wanted, representing one gene regulation network.

In order to combine the two informations, we simply ranked all edges $k \rightarrow l$ by the sum of their weights $w^m_{kl} + w^e_{kl}$. This ranking is denoted in the rest of the paper as the "genes-and-markers" ranking.

To further evaluate the influence of the post-processing on the ultimate results and the specific contribution of the causal marker-expression information, we also tested two additional post-processings. The first variant focuses on the genetic information and ranks each edge $k \rightarrow l$ using the w^m_{kl} weight only. Whenever marker j influences the expression of gene g, the strong correlations between the allelic states of j and other close markers on the genome (caused by genetic linkage) may easily lead to spurious predictions involving neighboring markers. We therefore scanned all markers along the genome and removed all edges that were locally dominated by other edges with the same target. More precisely, for all genes g, we removed any marker i such that there exists another marker j within a window of five markers containing i with $w^m_{jg} > w^m_{ig}$. The resulting ranked list of edges is called the "filtered markers" ranking in the following sections.

To evaluate the importance of this filtering process, we also included this marker filtering process in the initial "genes-and-markers" ranking method. In this post-processing method, the w^e and the w^m weights are combined by addition, as in the initial "genes-and-markers" post-processing, but only weights w^m coming from the list of "filtered-markers" edges are used. This third post-processing is naturally termed "genes-and-filtered-markers."

2.3 Results

2.3.1 General Analysis of the Predicted Networks

The above methods have been applied to the 72 datasets, including cases with 100, 1,000, and 5,000 genes. In this section we report results on the 1,000 gene datasets. The 100-gene networks were essentially provided as test datasets. The general trends in the 5,000 gene situation is that it is similar to the 1,000 gene situation (except for the fact that the Bayesian network approach could not be applied, our current implementation being limited to 4 GB corresponding to less than 3,000 variables, providing less opportunity for comparison) with slightly degraded overall performances.

We first present in Table 2.1 performances obtained by individual approaches as the area under the precision versus recall curve (AUPR). We remind the reader that

Table 2.1 Area (in percentage) under the precision versus recall curve (AUPR) for predicted networks with 1,000 genes by, respectively, the Lasso penalized regression, the Dantzig selector, random forests (RF), and Bayesian networks (BN)

Network/configuration/ data-set	AUPR with edge orientations				AUPR without edge orientations			
	Methods				Methods			
	Lasso	Dantzig	RF	BN	Lasso	Dantzig	RF	BN
Net4-Conf1-DS25-300SH	11.65	12.02	9.63	**14.20**	15.76	**16.41**	11.06	15.90
Net4-Conf2-DS26-900SH	15.88	15.66	17.95	**18.30**	**21.97**	21.68	20.05	20.08
Net4-Conf3-DS27-300SL	11.20	11.35	3.88	**11.83**	16.64	**17.18**	5.29	15.01
Net4-Conf4-DS28-900SL	21.49	21.78	9.64	**27.28**	32.46	**33.30**	11.31	32.95
Net4-Conf5-DS29-300DH	4.89	5.02	**7.31**	7.13	6.97	7.29	8.41	**8.60**
Net4-Conf6-DS30-900DH	9.68	10.05	13.82	**20.15**	13.81	14.53	15.60	**22.23**
Net4-Conf7-DS31-300DL	8.60	9.57	3.09	**13.18**	13.07	14.95	4.38	**16.59**
Net4-Conf8-DS32-900DL	16.20	17.43	7.39	**23.24**	24.20	26.71	9.12	**28.76**
Net5-Conf1-DS33-300SH	16.05	15.71	16.16	**16.96**	**21.52**	21.27	17.81	18.89
Net5-Conf2-DS34-900SH	22.17	21.71	23.96	**30.46**	31.08	30.64	26.28	**32.25**
Net5-Conf3-DS35-300SL	14.55	**14.61**	5.56	13.28	21.69	**22.10**	7.42	16.89
Net5-Conf4-DS36-900SL	24.57	24.70	13.53	**25.56**	37.38	**37.85**	15.86	31.37
Net5-Conf5-DS37-300DH	6.66	6.74	**9.04**	8.71	9.34	9.63	**10.58**	10.27
Net5-Conf6-DS38-900DH	12.80	12.67	21.76	**23.74**	17.55	17.76	23.73	**25.66**
Net5-Conf7-DS39-300DL	10.71	11.16	3.60	**15.36**	17.10	18.19	5.20	**18.71**
Net5-Conf8-DS40-900DL	17.42	17.92	11.04	**25.57**	26.33	27.75	12.86	**30.71**
Net6-Conf1-DS41-300SH	13.07	12.83	13.34	**15.75**	**17.90**	17.64	15.05	17.72
Net6-Conf2-DS42-900SH	17.54	17.59	23.63	**24.13**	24.81	24.80	25.56	**26.14**
Net6-Conf3-DS43-300SL	12.62	12.72	4.32	**13.40**	19.00	**19.38**	5.64	17.02
Net6-Conf4-DS44-900SL	20.72	**21.07**	10.67	20.14	32.06	**32.72**	12.69	26.12
Net6-Conf5-DS45-300DH	5.43	5.51	**7.41**	5.70	7.79	7.98	**8.83**	6.98
Net6-Conf6-DS46-900DH	8.55	8.43	**15.90**	12.34	11.91	11.95	**17.67**	14.13
Net6-Conf7-DS47-300DL	8.70	9.23	2.57	**10.07**	13.69	**14.84**	3.98	13.42
Net6-Conf8-DS48-900DL	14.68	15.33	7.82	**16.11**	22.86	**24.41**	10.06	21.36

Each network name ends by a short string which defines the sample size ($n = 300$ or 900), the gene density (D/S for dense or sparse), and the simulated gene expression heritability (H/L for high or low)

the recall is the ratio of correctly predicted edges among all edges to predict. The precision is the ratio of correctly predicted edges among all predicted edges. Since the predicted list of edges is ranked, edges are successively introduced with decreasing confidence scores, and precision and recall levels are computed at each step, defining a curve in the precision–recall space. The AUPR score is a compromise of the global performance of the method. It is a usual criterion in the Machine Learning community. Its values range from 1 (perfect recovery of all true edges with no error) to 0 (all predicted edges are incorrect). Note that a simple random guessing of edges should produce an AUPR of $\frac{\sharp\{\text{true edges}\}}{\sharp\{\text{possible edges}\}}$ often close to 0 if the network is sparse (with a number of edges much lower than the potential number of directed edges of $n(n-1)$). We report AUPR scores with and without taking edge orientations into account.

Sensitivity to Simulated Parameters

As described in Chap. 1, the different datasets were generated by varying different parameters: the underlying directed network, the number of simulated RILs (samples), the spacing between markers (and therefore gene density given the one gene/one marker assumption), and the gene expression heritability. The first observation that can be done on these datasets is the sensitivity of all methods to sample size that was expected. For 300 individuals (odd configuration numbers), the AUPR is usually between 5 and 15 % roughly. These results are weaker than the results presented in Vignes et al. (2011), using a set of different simulation parameters. With 900 individuals, the performance of all methods almost doubles, but still remains relatively poor.

Considering the inter-marker distance (low distance for configurations 5–8), all methods were sensitive to gene density. The higher the gene density, the lower the AUPR for the linear regressions and the random forest-based method. Indeed, with a high density marker map, the allelic states of two neighboring markers are correlated. It therefore becomes more difficult to precisely predict which marker i regulates a gene g: all neighboring markers offer essentially the same predictive information. Compared to these methods, the Bayesian network approach benefits from the dedicated marker pre-processing aimed at identifying *cis*-regulatory markers. When such a *cis* marker is detected, the corresponding edge is forced into the Bayesian network structure, avoiding increasingly likely mistakes in high-density configurations.

This marker distance effect became less important in configurations with a great number of individuals. This is due to the increased power and reliability of the regulatory marker localization and the expression data itself that becomes more informative.

From a more usual linkage analysis point of view, we must point out that the evaluation criteria used here is a very hard one. Imagine M_i is the regulatory marker that should have been identified as influencing gene g. If M_{i+1} (or any other neighbor marker/gene) is predicted instead, then the difference in terms of the chromosomal region that influences E_g is negligible and becomes increasingly small with an increasing gene density. Despite this increasingly small error, our edge detection criteria is purely Boolean and counts any neighbor prediction as a totally bad prediction. This probably advocates for new evaluation criteria beyond a pure $0/1$ edge detection event.

Another possibility that could explain this sensitivity to inter-marker distances lies in the quality of our confidence score. Compared to Vignes et al. (2011), the explored grid of penalizations was reduced to only $q = 10$ different values (instead of 20), offering a coarser image of the frequency of inclusion of a variable. This was required, from a computational point of view, because of the additional computational burden generated by bootstrapping. Another possibility would be that the bootstraping itself facilitates the selection of neighboring markers, because of the generated sampling noise. Our later evaluation of the bootstrapping influence (in one network), however, tends to show this is probably not the case.

Finally, all methods seem to be sensitive to the simulated gene expression heritability. In many cases, and quite unexpectedly at first, a weak heritability (high biological variance) seems to be favorable, especially for linear regression approaches. A likely explanation lies in the relationships that one can observe between the expression levels of two genes g_1 and g_2 that have a common parent regulator. In high heritability situations, a strong nonlinear dependency between the two genes was visible. In low heritability situations, the nonlinearity of the relationship seemed to fade away, which was favorable to linear regression methods. A possible way to enhance the behavior of linear regression in these conditions would be to apply a power transform such as the Box–Cox transform (Box and Cox 1964), a useful data pre-processing technique used to stabilize variance and make the data more normal distribution-like.

Comparison of Different Methods

The Lasso regression and the Dantzig selector had, as expected, very similar performances (both are linear regression approaches, using the same underlying model). Although, Dantzig seemed to offer better performances than Lasso in a majority of cases, the difference was usually negligible. This is different from our previous comparison in Vignes et al. (2011), where the Dantzig selector seemed to offer better performance than the Lasso regression (the Dantzig selector approach finished second of the DREAM5 challenge, and was outperformed only by a consensus meta-analysis that exploited its predictions). More tests would be needed to check whether this could be caused by the shift in the underlying optimisation method. Vignes et al. (2011) used an exact simplex-based linear programming solver, while we used a more efficient homotopy-based method (Asif and Romberg 2010) to cope with the additional computational burden generated by the bootstrapping process.

Surprisingly, our random forest approach often gave the worst performance (except on configurations 2 and 6, with high heritability). A likely explanation for the limited performance of our random forest approach lies in the fact that we handled expression data E_i and allelic states M_j together, as possible splitting variables in the same trees. However, these two variables are very different. Expression data is a continuous variable, which offers a lot of freedom on possible splitting decisions. Allelic states are Boolean variables, and therefore offer no freedom in the possible splitting decision. The criteria used to grow forests and decide which variable is taken next being its ability to reduce the variance in the separated datasets, it is likely that allelic state variables tend to be inserted lately compared to expression data variables. This tendency is confirmed by Strobl et al. (2007) when random forests are used for classification purpose: the algorithm tends to be biased and to be more likely to select variables with a higher number of levels than variables with few levels like binary variables. This may lower the performance of the approach a lot compared to others. A better approach would probably be to handle expression data variables and allelic state variables in two independent random forest constructions to later mix the associated confidence scores in a single confidence score (Geurts and Huynh-Thu 2012) instead of mixing the simultaneous edge ranks as we did (see also

Chap. 7 of this book). The better performances (compared to other methods) on the configurations 2 and 6 (high heritability) can also be explained with the previously mentioned highly nonlinear relationships that appear between co-regulated genes in this case. The ability of random forests to capture nonlinear relationships may explain its relative success on these configurations.

Overall, Bayesian networks offer the best performance in most cases, comparable to those obtained in Vignes et al. (2011). We observe that each method has an advantage in a specific area of the parameter space, showing their potential complementarity.

Directed or Undirected Edges

In our linear regression methods, the current post-processing only partially allows us to orientate edges, thanks to marker data: it is known that genetic content influences observed phenotypes, including gene expression levels, and not the converse. Learnt relationships between variables that account for gene expression levels are symetrical, hence the predicted direct links between these variables are given a disadvantage. The other way to see this is that edges from a marker variable to a gene expression are favored even if they are slightly less supported by the data according to a modeling criterion. This may result in either spurious marker to gene expression arcs being assigned relative scores higher than they should or significant relationships between gene expression levels being moved back in the list of predicted edges. For this reason, we also evaluate the performances considering undirected edges. This analyzes the ability to predict noncausal relationships between genes. The corresponding AUPR scores are given in the four rightmost columns of Table 2.1.

The limited ability of penalized linear regression methods to infer causality (which is only done when a polymorphism is detected as a possible regressor of the expression level) leads to visible improvements in their predictive capabilities. The Lasso regression and the Dantzig selector are performing best in only two cases if directed edges are considered and in 12 cases over 24 in the undirected edge case. In the best cases, the Dantzig selector can offer an AUPR of almost 40 %, with performances that more frequently exceed those of Bayesian networks.

Overall, both the Bayesian Network approach and the random forest approach seems to benefit less from this relaxation in the evaluation criteria. This means that when they predict a directed edge, in most cases, this edge may be a correct directed edge or else, it is often a false edge (even ignoring orientation).

Table 2.2 Effect of different post-processings on the AUPR scores (in percentage) for learning Net5-Conf2-DS34 ($n = 900$, sparse marker map and high gene expression heritability) with the Lasso regression, the Dantzig selector and our random forest method (RF)

Post-processing	Methods		
	Lasso	Dantzig	RF
Genes-and-markers	22.17	21.71	23.96
Genes-and-filtered-markers	23.38	23.21	**25.66**
Filtered-markers	**28.10**	**26.30**	20.05

2.3.2 *Focus on a 1,000 Gene Network with* $n = 900$*, High Gene Expression Heritability, and 5 cM Between Consecutive Markers (Net5-Conf2-DS34-900SH)*

Implementing a practical learning scheme for GRN in System Biology cannot be just characterized by the underlying mathematical method used to perform the inference itself. The modeling, the pre- and post-processing of the data, possible bootstrapping (and number of bootstrap samples), the criteria used for evaluation may all have non-negligible impacts on the final results.

2.3.2.1 Post-processing

To analyze the importance of the post-processing used on the obtained results, we compared two other post-processing methods on one network. We used dataset 34. With a large sample, a sparse gene density and high heritability, it defines a playground where most methods (including our random forest approach) obtain comparable performances for the directed edges prediction. The Bayesian network approach is excluded from this comparison since it uses a specific pre-processing that could hinder the effect of the different post-processings.

Table 2.2 gives the AUPR obtained using the previous "genes-and-markers" post-processing (which combines expression data-based scores w^e and allelic states-based scores w^m by addition) with two extra post-processing (described in Sect. 2.2.6). The "filtered-markers" post-processing focuses on marker information by using only w^m scores. It also tries to weaken the influence of high correlations between close markers by keeping only markers with local undominated w^m score. The "genes-and-filtered-markers" post-processing combines this filtering process in the original "gene-and-markers" post-processing.

On this dataset, one can check that a change in the post-processing may change the ranking of the different methods. On dataset 34, retaining only allelic state-based scores gave much better results for linear regression methods but was rather disappointing for random forests. As we mentioned previously, the Boolean allelic state variables are probably introduced lately in the regression trees, giving priority to expression levels. This may explain the important effect of keeping only allelic state-based scores on random forests. The mixed "genes-and-filtered-markers" post-processing gave more even performances.

All three post-processing were tested on all configurations of the whole datasets. The "filtered-markers" post-processing gave the overall best result on dataset 34, but it gave extremely poor results on most other configurations (except for configurations 1 and 2, all AUPR being below 11 %). The "genes-and-filtered-markers" post-processing often offered performances that were comparable to our initial "genes-and-marker" post-processing, but was more often dominated by it than the converse.

Clearly, because of the dual nature of genetical genomics data that captures different effects through different types of variables, there are plenty of possible choices for combining the two types of information. A sensible conclusion at this point is that the optimal post-processing depends on features of the data that need to be carefully checked before proposing analysis results.

2.3.2.2 The AUPR Curve

To be consistent with the previous DREAM5 experiment, we used the AUPR criteria to compare the performances of the different approaches. In this section, we want to show that AUPR is a very high-level criterion. It summarizes the performance of every method as a unique number, but hides different behaviors.

In Fig. 2.1, we give the complete precision versus recall curve for all methods on dataset 34 using two different post-processings ("genes-and-markers" and the best post-processing from Table 2.2). Some methods (such as the Dantzig selector in the "filtered-markers" post-processing) tend to have a very high precision initially. This means that almost all of the first few hundred predicted edges are correct. On this same dataset, the Bayesian network learning presented a similar AUPR but a very different behavior. The initial precision decreased more rapidly. This means that false positive edges existed in the beginning of the list of ranked edges, making the output harder to use. Again, this comparison between the Dantzig selector and the Bayesian network learning is valid for this dataset and this post-processing but conclusion may differ on another dataset and with different post-processings.

2.3.2.3 Effect of the Chosen Number of Bootstraps

The use of bootstrapping is quite costly in terms of computation time: all cpu times are immediately multiplied by our $N_{\text{boot}} = 100$ bootstraps. We checked whether this additional work is of interest first beyond theoretical grounds and whether it is sufficient. We therefore compared the results obtained using bootstrapping with an increasing number of bootstraps. The results obtained on dataset 34 are presented in Fig. 2.2 using the Lasso regression approach and two different post-processings. These curves give the observed recall at different precision levels. Precision levels

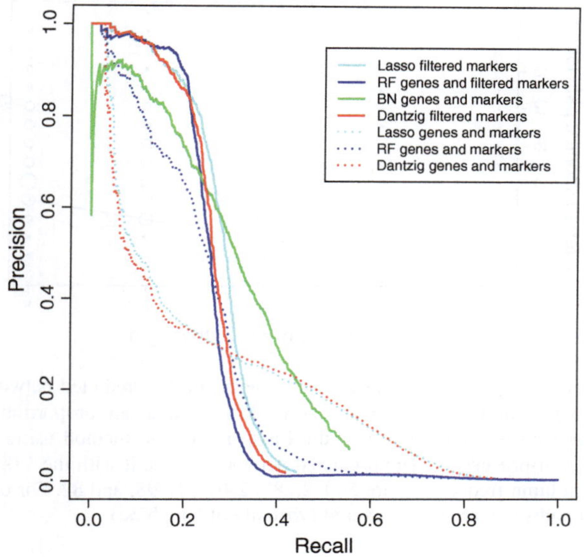

Fig. 2.1 PR *curve* for the different methods and different post-processings on dataset Net5-Conf2-DS34-900SH

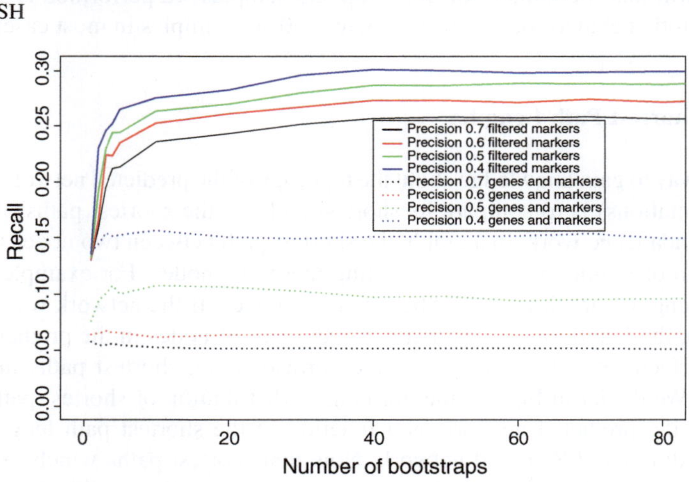

Fig. 2.2 Effect of the number of bootstrapped samples on dataset Net5-Conf2-DS34-900SH for different levels of precision with the Lasso regression method using the "genes-and-markers" and the "genes-and-filtered-markers" post-processings

range from a nearly acceptable degree of precision of 0.4 to a good degree of precision of 0.7. One can see that bootstrapping enhances performance, but the importance of the improvement varies a lot depending on the post-processing used. These curves

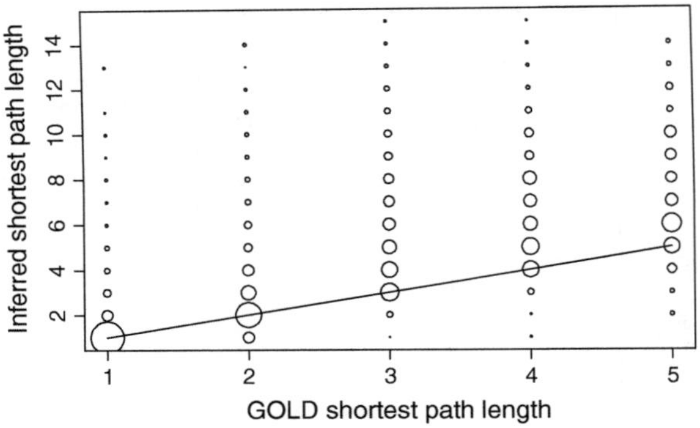

Fig. 2.3 Normalized frequency for shortest path lengths in the predicted network as a function of shortest path lengths in the Gold Standard network. Disc areas are proportional to normalized frequencies. Dataset 34 was analyzed with the Lasso regression method using the "genes-and-filtered-markers" post-processing. The predicted network was built with the 1,000 first predicted edges. Raw total column frequencies are 540, 2,283, 2,407, 1,098, and 82. For clarity, we do not represent shortest paths that are not recovered (virtually of length ∞)

also confirm that the number of bootstrap sub-samples we performed is sufficient, an asymptotic behavior being reached before 50 sub-samples in most cases.

2.3.2.4 Shortest Path Lengths

Another way to gain knowledge about the topology of the predicted network (without edge orientations) is to compare its shortest paths to the shortest paths of the true (Gold Standard) network. In a graph, the shortest path between two nodes is defined as the path of minimum length connecting these two nodes. For example, shortest paths of length 1 are direct edges between two nodes. If the network was perfectly recovered, the length of shortest paths for any pair of nodes in the predicted graph would perfectly match the length of the corresponding shortest paths in the true network. We depict in Fig. 2.3 the normalized distribution of shortest path lengths in one of the predicted network as a function of the shortest path lengths in the corresponding Gold Standard network. Note that shortest paths which exist in the true network but whose extremities are not connected in the predicted network are not accounted for here. The rationale here is to check whether the distribution of distances between genes of the two compared networks are close. If this is the case, the distribution should concentrate around the $y = x$ axis.

For example, we found 2,341 shortest paths of length one (i.e., edges) that were not connected at all in the predicted network. It means that among the 2,882 true edges to recover, 439 (implying a recall of 15.2 %) were correctly identified by the predicted network while 102 were found with at least another intermediate gene in the path between them, and 2,341 pair of genes were not connected in the predicted network. This is due to the fact that the predicted network was not fully connected, and that only 1,000 edges were predicted while the true network contained three times more edges.

Focusing on the first two columns of Fig. 2.3, we see that the predicted graph had a distribution of shortest paths of lengths 1 and 2 that matched with the actual distribution of such edges in the true network. This was not the case anymore for paths of length 3 and over: the predicted network tended to overestimate the length of such shortest paths when it was able to identify them. Again, the fact that we only studied the first 1,000 predicted edges plays an important role in this observation. Longer paths are needed to cope with correct paths of a given length that are composed of true edges only. However, even those few edges and some of them that were not correct (the precision level was here below 0.5), the distributions of shortest path lengths in the predicted network and in the true network were not too different.

2.4 Conclusion

In contrast with our previous experiments in Vignes et al. (2011), which relied on a subset of the same algorithms (the Lasso regression, the Dantzig selector and Bayesian network structure learning) on data generated using the same generator but with simplified settings, the results we obtained here with additional effort (computational cost of the bootstrapping process and integration of a random forest approach) are rather disappointing in the hardest situations, in terms of AUPRs. On these hardest problems, whether because of limited number of individuals, or because of nonlinear relationships induced by high heritability, these low AUPR often hide rapidly decreasing precision, which means that the list of predicted edges quickly becomes hard to exploit to predict gene regulations.

Given that boostrapping seems to be able to provide improved prediction quality, the most likely reason of these results lies in the generated data itself which relies on new parameters and new networks. Clearly, the specific characteristics of these new datasets could probably be accommodated, to some extent, by the same methods, using different modeling or pre/post-processing strategies. In the case of configuration 2, the use of the "filtered-marker" post-processing strategy provided significant improvements in the AUPR of the linear regression-based approaches, leading to respectable performances.

These results point also to the ambitious aim of trying to identify which statistical method would perform best for GRN learning in a genetical genomics context, based on a large scope of methods and on large datasets of simulated data. The overwhelming size of combinations that can be imagined using different methods, models and pre and post-processing procedures to deal with a large amount of simulated data which is itself parameterized by several parameters defines a daunting task. It is quite obvious for us that we have only explored a small fraction of all possible combinations and that much work remains to be done to identify an ideal performer, if it exists. Indeed, our results show that every method, even those that tend to give the worst results, can outperform other methods in specific situations. A combination of methods may therefore be, ultimately, the best approach.

These experiments also show the possible difficulty in the generation of realistic simulated data for genetical genomics. Given the radically different results obtained depending on the change of parameters used for problem generation, it is natural to wonder which of these different settings could be considered as the most realistic model for real genetical genomics data and for which organism. First elements of response to this question may finally open the door to tackle real datasets.

Acknowledgments We are very grateful to the staff of the GenoToul (Toulouse, France) Bioinformatics plateform for the computational support it provided during this work. We would also like to thank our colleagues from CRS4 Bioinformatica for creating the datasets of the present study.

References

Asif MS, Romberg JK (2010) Dynamic updating for ℓ_1 minimization. J Sel Top Sig Process 4(2):421–434

Aten JE, Fuller TF, Lusis AJ, Horvath S (2008) Using genetic markers to orient the edges in quantitative trait networks: the NEO software. BMC Bioinform 2:34

Bach F (2008) Bolasso: model consistent lasso estimation through the bootstrap. In: Cohen WW, McCallum A, Roweis ST (eds) Proceedings of the twenty-fifth international conference on machine learning (ICML), ACM international conference proceeding series, vol 307. Helsinki, Finland, pp 25–32

Bansal M, di Bernardo D (2007) Inference of gene networks from temporal gene expression profiles. IET Syst Biol 1(5):306–312

Box GEP, Cox DR (1964) An analysis of transformations. J Roy Stat Soc Ser B (Methodological), 26(2):211–252

Breiman L (2001) Random forests. Mach Lear 45(1):5–32

Candès E, Tao T (2007) The Dantzig selector: Statistical estimation when p is much larger than n. Ann Stat 35(6):2313–2351

Chickering D, Heckerman D, Meek C (2004) Large-sample learning of Bayesian networks is NP-hard. J Mach Learn Res 5:1287–1330

Efron B, Tibshirani R (1997) Improvements on cross-validation: The. 632+ bootstrap method. J Am Stat Assoc 92(438):548–560

Efron Bradley (1981) Nonparametric estimates of standard errors - the jackknife, the bootstrap and other methods. Biometrika 68(3):589–599

Friedman J, Hastie T, Tibshirani R (2010) Regularization paths for generalized linear models. J Stat Softw 33(1):1–22

Friedman N, Nachman I, Peér D (1999) Learning bayesian network structure from massive datasets: The "sparse candidate" algorithm. In: Proceedings of the 15th conference on uncertainty in artificial intelligence, Stockholm, Sweden, pp 206–215

Friedman N, Linial M, Nachman I, Peer D (2000) Using Bayesian networks to analyse expression data. J Comput Biol 7(3):601–620

Geurts P, Huynh-Thu V-A (2012) Personal communication

Hastie T, Tibshirani R, Friedman J (2009) The elements of statistical learning: data mining, inference, and prediction. Series in Statistics, 2nd edn. Springer, New York

Heckerman D, Geiger D, Chickering DM (1995) Learning Bayesian networks: the combination of knowledge and statistical data. Mach Learn 20(3):197–243

Huynh-Thu VA, Irrthum A, Wehenkel L, Geurts P (2010) Inferring regulatory networks from expression data using tree-based methods. PLoS ONE 5(9):e12776

Jansen RC, Nap NP (2001) Genetical genomics : the added value from segregation. Trends Genet 17(7):388–391

Koller D, Friedman N (2009) Probabilistic graphical models: principles and techniques. MIT press, Cambridge

Lèbre S, Becq J, Devaux F, Stumpf MH, Lelandais G (2010) Statistical inference of the time-varying structure of gene-regulation networks. BMC Systems Biology 4:130

Leclerc RD (2008) Survival of the sparsest: robust gene networks are parsimonious. Mol Syst Biol 4:213

Liaw A, Wiener M (2002) Classification and regression by randomforest. R News 2(3):18–22

Liu B, de la Fuente A, Hoeschele I (2008) Gene network inference via structural equation modeling in genetical genomics experiments. Genetics 178(3):1763–1776

Marbach D, Mattiussi C, Floreano D (2009) Replaying the evolutionary tape: biomimetic reverse engineering of gene networks. Ann New York Acad Sci 1158(1):234–245

Pinna A, Soranzo N, Hoeschele I, de la Fuente A (2011) Simulating systems genetics data with SysGenSIM. Bioinformatics 27(17):2459–2462

Rau A, Jaffrezic F, Fouley J-L, Doerge RW (2010) An empirical Bayesian method for estimating biological networks from temporal microarray data. Stat Appl Genet Mol Biol 9(1):art.9

Strobl C, Boulesteix A-L, Zeileis A, Hothorn T (2007) Bias in random forest variable importance measures: Illustrations, sources and a solution. BMC Bioinform 8:25

Thomas R (1973) Boolean formalization of genetic control circuits. J Theor Biol 42(3):563–585

Tibshirani R (1996) Regression shrinkage and selection via the lasso. J Roy Stat Soc Ser B (Methodological), 58(1):267–288

Vandel J, Mangin B, de Givry S (2012) New local move operators for Bayesian network structure learning. In: Proceedings of PGM-12, Granada, Spain

Vignes M, Vandel J, Allouche D, Ramadan-Alban N, Cierco-Ayrolles C, Schiex T, Mangin B, de Givry S (2011) Gene regulatory network reconstruction using Bayesian networks, the Dantzig selector, the lasso and their meta-analysis. PloS one 6(12):e29165

Yvert G, Brem RB, Whittle J, Akey JM, Foss E, Smith EN, Mackelprang R, Kruglyak L (2003) Trans-acting regulatory variation in saccharomyces cerevisiae and the role of transcription factors. Nat Genet 35(1):57–64

Zhu J, Wiener MC, Zhang C, Fridman A, Minch E, Lum PY, Sachs JR, Schadt EE (2007) Increasing the power to detect causal associations by combining genotypic and expression data in segregating populations. PLoS Comput Biol 3(4):e69

Chapter 3
Benchmarking a Simple Yet Effective Approach for Inferring Gene Regulatory Networks from Systems Genetics Data

Sandra Heise, Robert J. Flassig and Steffen Klamt

Abstract We apply our recently proposed gene regulatory network (GRN) reconstruction framework for genetical genomics data to the StatSeq data. This method uses, in a first step, simple genotype–phenotype and phenotype–phenotype correlation measures to construct an initial GRN. This graph contains a high number of false positive edges that are reduced by (i) identifying eQTLs and by retaining only one candidate edge per eQTL, and (ii) by removing edges reflecting indirect effects by means of TRANSWESD, a transitive reduction approach. We discuss the general performance of our framework on the StatSeq in silico dataset by investigating the sensitivity of the two required threshold parameters and by analyzing the impact of certain network features (size, marker distance, and biological variance) on the reconstruction performance. Using selected examples, we also illustrate prominent sources of reconstruction errors. As expected, best results are obtained with large number of samples and larger marker distances. A less intuitive result is that significant (but not too large) biological variance can increase the reconstruction quality. Furthermore, a somewhat surprising finding was that the best performance (in terms of AUPR) could be found for networks of medium size (1,000 nodes), which we had expected to see for networks of small size (100 nodes).

3.1 Introduction

Systems Genetics approaches provide a new paradigm of large-scale genome and network analysis (Jansen and Nap 2001; Jansen 2003; Rockman and Kruglyak 2006; Rockman 2008). These methods use naturally occurring multifactorial perturbations (e.g., polymorphisms) to causally link genetic or chromosomal regions to observed

S. Heise · R. J. Flassig · S. Klamt(✉)
Max Planck Institute for Dynamics of Complex Technical Systems,
Sandtorstrasse 1, D-39106 Magdeburg, Germany
e-mail: klamt@mpi-magdeburg.mpg.de

A. de la Fuente (ed.), *Gene Network Inference*,
DOI: 10.1007/978-3-642-45161-4_3,

phenotypic trait data. Identifying a chromosomal region (the quantitative trait locus (QTL)) that influences a certain phenotypic trait is known as QTL mapping. In genetical genomics, a particular subclass of systems genetics, gene-expression levels are considered as phenotypic traits (called etraits) and identified QTLs are referred to as expression-QTLs (eQTLs). One application of eQTL maps obtained from genetical genomics approaches is the reconstruction of gene regulatory networks (GRNs).

According to Liu et al. (2010), a GRN reconstruction pipeline for genetical genomics data consists of three major steps: (i) eQTL mapping, (ii) candidate regulator selection, and (iii) network refinement. Step (i) is used to identify chromosomal regions (eQTLs) that impact on expression levels (=traits) of genes. A detailed review on eQTL mapping is, for instance, given by Michaelson et al. (2009). In step (ii), the eQTL map in combination with a genetic map is used to select single candidate (regulator) genes from the eQTLs. Frequently used methods include conditional correlation (Bing and Hoeschele 2005; Keurentjes et al. 2007), local regression (Liu et al. 2008), or analysis of between-strains SNPs (Li et al. 2005). In the third step (iii), network refinement methods are employed to the topology obtained in step (ii), e.g., with the goal to identify and eliminate (false positive) edges arising from indirect effects. Here, Bayesian network approaches (Zhu et al. 2007) and structural equation modeling, SEM, (Liu et al. 2008) have been used.

In this chapter, we apply our recently proposed GRN reconstruction framework for genetical genomics data (Flassig et al. 2013), which incorporates the three major reconstruction steps mentioned above in a modular fashion. The framework follows a *simple-yet-effective* paradigm: it is based on simple correlation measures, without the need for computational demanding optimization steps. This approach is therefore suited for small- and large-scale networks and performed comparable well in the case of little samples but many genes, as we illustrate in Flassig et al. (2013) using simulated and biological data. The workflow of the framework is shown in Fig. 3.1. The initial GRN is constructed based on genotype–phenotype and phenotype–phenotype correlation analyses. Due to genetic linkage there are often groups of genetically adjacent regulator gene candidates, which target the same gene resulting into eQTLs. To avoid many false-positive interaction predictions, single candidate regulators are therefore identified from the eQTLs. Finally, as a method for network refinement in step (iii), indirect path effects are removed by TRANSWESD, a transitive reduction approach introduced recently (Klamt et al. 2010).

3.2 Methods

Figure 3.1 shows the general workflow of our reconstruction framework together with a simple illustrative example. Starting from a typical set of genetical genomics data that include genotyped markers, phenotyped genes and gene-to-marker association, marker linkage analysis, and genotype assignment for each gene is performed in a preprocessing step. In particular, a linkage map is generated in which two markers are indicated to be genetically linked if their genotype–genotype correlation

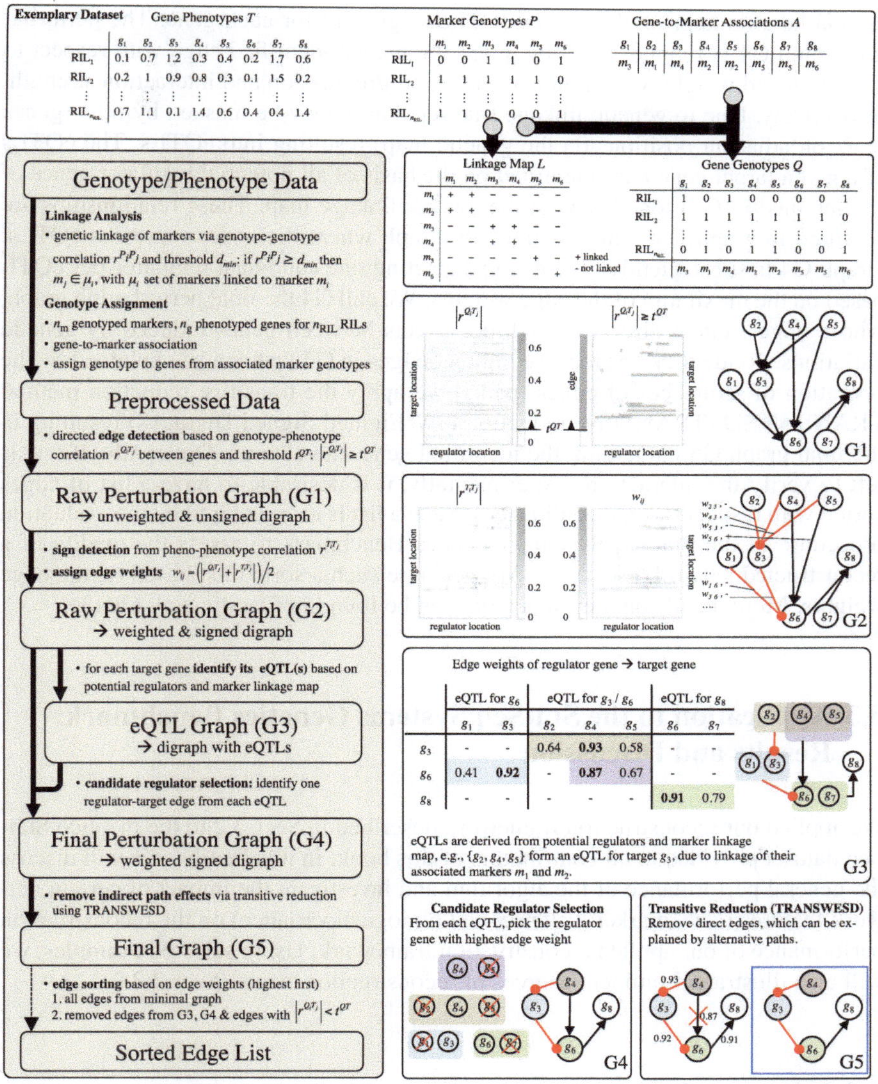

Fig. 3.1 Workflow of the proposed framework for reconstructing GRNs from genetical genomics data (*left*) with an illustrative example (*top panel* and *right*). For detailed explanations see text. Reproduced with permission of Oxford University Press from Flassig et al. (2013)

exceeds a given threshold parameter $d_{\min}$. Then, in a first step, an unweighted and unsigned perturbation graph G1 is derived in which an edge $i \rightarrow j$ is included if their corresponding genotype-phenotype correlation exceeds a second threshold t^{QT}. The nodes in the graph directly correspond to genes while the linkage map (of the markers) is kept to allow later eQTL assignment for each gene. The perturbation graph G1 is refined to G2 by quantifying each identified edge with respect to edge sign and weight, which indicate activation/repression and interaction strength, respectively. Due to genetic linkage true regulators may be masked by other genes (e.g., on adjacent positions on the genetic map) resulting into eQTLs. The eQTLs of a given target gene t are identified on the basis of all potential regulator genes of t (contained in G2) together with the marker linkage map. These relationships are captured in graph G3, which is the only graph where the nodes represent eQTLs. Graph G4 is subsequently obtained by selecting one candidate regulator per eQTL based on the maximum of the edge weights. We call G4 the final perturbation graph, whose edges reflect direct and indirect effects between genes induced by genetic variations. To identify and remove indirect edges in G4 that can be explained by the operation of sequences of edges (paths) we apply the transitive reduction method TRANSWESD (TRANSitive reduction in WEighted Signed Digraphs) resulting in the final graph G5 containing the identified gene interactions. Optionally, if one is left to verify the interactions experimentally, it is desirable to have a list of edges sorted with respect to edge confidences. Such a list is also required by the evaluation procedure of the StatSeq Systems Genetics Benchmark to assess the quality of a reconstructed network (Sect. 3.3). We generate such a sorted list based on the edge weights. More details on the framework can be found in Flassig et al. (2013).

3.3 Application to the StatSeq Systems Genetics Benchmark: Results and Discussion

We applied our reconstruction framework described in Sect. 3.2 to the in silico StatSeq dataset provided to all contributors of this book. In this section, we will discuss the general performance of the algorithm and investigate the impact of certain network features (size, marker distance, and biological variance) on the reconstruction performance of our applied reconstruction framework. Using selected examples, we will also illustrate prominent sources of reconstruction errors (Sect. 3.3.2).

3.3.1 General Performance Analysis with Respect to Network Configurations

Table 3.1 shows the AUPR and AUROC reconstruction performance (obtained by using optimal values for the thresholds $d_{\min}$ and t^{QT}) for all studied 72 network configurations: 3 different network sizes (100, 1000, 5000) $\times$ 3 replicates (with same topological parameters) $\times$ 2 marker distances (close and far) $\times$ 2 different biological variances (high and low) $\times$ 2 different population sizes (300 and 900) (see also Chap. 1). The performance measures are given for graph G2, G4, and G5

Table 3.1 Reconstruction performance obtained for each network configuration achieved with the indicated optimal parameter values

Configuration	t^{QT}	$d_{\min}$	G2				G4				G5			
			AUROC	AUPvR	TP	FP	AUROC	AUPvR	TP	FP	AUROC	AUPvR	TP	FP
[100.1.1]: F / L / 300	0.26	0.5	0.7810	0.2247	104	516	0.7842	0.2782	69	38	0.7843	0.2801	69	36
[100.1.2]: F / L / 900	0.22	0.6	0.8058	0.2007	129	669	0.8087	0.2772	81	64	0.8082	0.2750	79	57
[100.1.3]: F / H / 300	0.14	0.2	0.8011	0.2178	97	771	0.8143	0.2527	67	125	0.8145	0.2548	67	118
[100.1.4]: F / H / 900	0.1	0.5	0.8370	0.3041	125	834	0.8491	0.3588	92	84	0.8494	0.3635	92	72
[100.1.5]: C / L / 300	0.6	1	0.7553	0.1043	27	111	0.7552	0.1216	25	83	0.7552	0.1219	25	81
[100.1.6]: C / L / 900	0.24	0.1	0.7561	0.0942	126	1706	0.7668	0.1253	35	61	0.7666	0.1245	34	59
[100.1.7]: C / H / 300	0.16	0.3	0.7792	0.1977	120	1596	0.8034	0.2593	62	69	0.8036	0.2611	62	65
[100.1.8]: C / H / 900	0.14	0.2	0.8062	0.2197	119	1529	0.8362	0.3104	67	41	0.8363	0.3111	67	39
[100.2.1]: F / L / 300	0.2	0.2	0.7675	0.2467	139	778	0.7644	0.2898	85	69	0.7647	0.3016	82	54
[100.2.2]: F / L / 900	0.16	0.5	0.7858	0.2050	179	1202	0.7722	0.2910	98	86	0.7727	0.3208	97	59
[100.2.3]: F / H / 300	0.12	0.2	0.7626	0.2198	110	1017	0.7781	0.2481	78	227	0.7791	0.2708	77	183
[100.2.4]: F / H / 900	0.12	0.5	0.8043	0.2972	124	563	0.8104	0.3432	96	58	0.8106	0.3613	92	43
[100.2.5]: C / L / 300	0.22	0.2	0.7307	0.1529	146	1821	0.7257	0.1902	54	81	0.7256	0.1914	53	71
[100.2.6]: C / L / 900	0.12	0.7	0.7355	0.1333	214	3386	0.7134	0.1831	61	142	0.7142	0.1939	61	115
[100.2.7]: C / H / 300	0.02	0.2	0.7524	0.2005	285	7953	0.7624	0.2430	92	322	0.7617	0.2379	80	220
[100.2.8]: C / H / 900	0.1	0.6	0.7686	0.2144	152	2104	0.7987	0.2965	65	71	0.7988	0.3038	65	64
[100.3.1]: F / L / 300	0.24	0.2	0.7725	0.2324	92	508	0.7755	0.2737	68	42	0.7756	0.2772	68	35
[100.3.2]: F / L / 900	0.2	0.4	0.7834	0.2505	127	821	0.7897	0.2886	84	68	0.7898	0.3013	83	49
[100.3.3]: F / H / 300	0.12	0.3	0.7456	0.1914	105	948	0.7596	0.2266	83	250	0.7613	0.2319	82	187
[100.3.4]: F / H / 900	0.14	0.6	0.7799	0.2479	86	430	0.7878	0.2901	74	74	0.7864	0.2884	73	72
[100.3.5]: C / L / 300	0.28	0.2	0.7124	0.1084	102	1043	0.7169	0.1566	46	44	0.7169	0.1579	46	43
[100.3.6]: C / L / 900	0.14	0.7	0.7156	0.1108	184	3068	0.7233	0.1676	65	135	0.7202	0.1642	60	119
[100.3.7]: C / H / 300	0.22	0.7	0.7400	0.1188	39	522	0.7533	0.1420	23	28	0.7533	0.1420	23	28
[100.3.8]: C / H / 900	0.24	0.2	0.7359	0.1190	51	463	0.7453	0.1612	27	11	0.7453	0.1612	27	11
Configuration average for 300 samples			0.7584	0.1846	114	1465	0.7661	0.2235	63	115	0.7663	0.2274	61	93
Configuration averaged for 900 samples			0.7762	0.1997	135	1398	0.7835	0.2577	70	75	0.7832	0.2641	69	63

(continued)

Table 3.1 (continued)

Configuration	t^{QT}	$d_{\min}$	G2				G4				G5			
			AUROC	AUPvR	TP	FP	AUROC	AUPvR	TP	FP	AUROC	AUPvR	TP	FP
[1000.1.1]: F / L / 300	0.15	0.4	0.8273	0.1654	1464	26333	0.8224	0.2037	1115	8658	0.8203	0.2060	998	4523
[1000.1.2]: F / L / 900	0.1	0.3	0.8725	0.2170	1887	28841	0.8647	0.3015	1397	4666	0.8612	0.3171	1275	2938
[1000.1.3]: F / H / 300	0.1	0.2	0.8354	0.1996	1309	91968	0.8499	0.2143	1082	33130	0.8567	0.2230	948	9124
[1000.1.4]: F / H / 900	0.1	0.2	0.8918	0.2842	1127	11114	0.8939	0.3417	1061	2756	0.8938	0.3430	1048	2531
[1000.1.5]: C / L / 300	0.15	0.5	0.8162	0.0872	1579	71242	0.8133	0.1479	748	6271	0.8130	0.1505	694	3588
[1000.1.6]: C / L / 900	0.1	0.2	0.8429	0.0992	2042	106265	0.8243	0.1774	985	5282	0.8218	0.1857	875	2716
[1000.1.7]: C / H / 300	0.1	0.3	0.8136	0.1431	1366	123310	0.8363	0.1773	892	16739	0.8395	0.1846	816	6011
[1000.1.8]: C / H / 900	0.1	0.2	0.8853	0.2094	1239	34817	0.8928	0.3185	1004	2351	0.8927	0.3226	998	2083
[1000.2.1]: F / L / 300	0.15	0.3	0.8676	0.2738	1513	24947	0.8670	0.3463	1241	8061	0.8667	0.3529	1171	4415
[1000.2.2]: F / L / 900	0.1	0.4	0.8976	0.3050	1844	29917	0.8969	0.3801	1513	4978	0.8952	0.4199	1420	2981
[1000.2.3]: F / H / 300	0.1	0.5	0.8554	0.2231	1221	92487	0.8679	0.2265	1076	46949	0.8804	0.2404	970	9557
[1000.2.4]: F / H / 900	0.1	0.3	0.9166	0.3379	1166	12491	0.9193	0.3743	1094	2960	0.9192	0.3854	1085	2658
[1000.2.5]: C / L / 300	0.15	0.4	0.8392	0.1114	1483	69014	0.8407	0.1747	813	5852	0.8407	0.1805	758	3342
[1000.2.6]: C / L / 900	0.15	0.4	0.8763	0.1541	1446	52275	0.8815	0.2718	964	1506	0.8813	0.2801	949	1180
[1000.2.7]: C / H / 300	0.1	0.3	0.8440	0.1592	1300	121251	0.8706	0.1914	900	16997	0.8743	0.2006	847	6211
[1000.2.8]: C / H / 900	0.1	0.2	0.8994	0.2107	1205	40136	0.9084	0.2993	950	2462	0.9085	0.3117	947	2160
[1000.3.1]: F / L / 300	0.2	0.5	0.8566	0.2079	1089	9353	0.8571	0.2716	966	1437	0.8571	0.2808	955	1210
[1000.3.2]: F / L / 900	0.15	0.4	0.8834	0.2195	1451	14163	0.8828	0.2871	1121	1372	0.8828	0.3109	1107	874
[1000.3.3]: F / H / 300	0.1	0.4	0.8398	0.2135	1366	94845	0.8515	0.2224	1183	41839	0.8605	0.2337	1047	9572
[1000.3.4]: F / H / 900	0.1	0.3	0.8948	0.2602	1144	11561	0.8969	0.2977	1054	2818	0.8969	0.3059	1048	2645
[1000.3.5]: C / L / 300	0.2	0.4	0.8246	0.0789	1241	38766	0.8240	0.1001	517	1881	0.8240	0.1017	502	1684
[1000.3.6]: C / L / 900	0.1	0.2	0.8613	0.0935	1941	95880	0.8574	0.1246	807	4475	0.8567	0.1293	728	2707
[1000.3.7]: C / H / 300	0.1	0.3	0.8243	0.1315	1338	122280	0.8481	0.1588	902	17016	0.8515	0.1680	824	6328
[1000.3.8]: C / H / 900	0.1	0.3	0.8844	0.1617	1232	38206	0.8927	0.2197	906	2407	0.8927	0.2252	899	2231
Configuration averaged for 300 samples			0.8370	0.1662	1356	73816	0.8457	0.2029	953	17069	0.8487	0.2102	878	5464
Configuration averaged for 900 samples			0.8839	0.2127	1477	39639	0.8843	0.2828	1071	3169	0.8836	0.2947	1032	2309

(continued)

Table 3.1 (continued)

Configuration	t^{QT}	d_{min}	G2				G4				G5			
			AUROC	AUPvR	TP	FP	AUROC	AUPvR	TP	FP	AUROC	AUPvR	TP	FP
[5000.1.1]: F / L / 300	0.25	0.6	0.8630	0.1839	4470	29115	0.8628	0.2226	934	4287	0.8628	0.2145	931	3781
[5000.1.2]: F / L / 900	0.15	0.4	0.9117	0.2672	6990	71653	0.9114	0.3865	2269	4802	0.9113	0.4022	2247	2688
[5000.1.3]: F / H / 300	0.2	0.2	0.8746	0.1451	2196	20255	0.8747	0.1480	192	8890	0.8747	0.1479	192	8863
[5000.1.4]: F / H / 900	0.1	0.3	0.9210	0.2665	5790	112803	0.9216	0.3004	1550	50216	0.9216	0.3030	1539	46881
[5000.1.5]: C / L / 300	0.2	0.3	0.8652	0.0570	5770	267008	0.8640	0.0910	1115	13438	0.8640	0.0924	1106	12350
[5000.1.6]: C / L / 900	0.15	0.4	0.8976	0.0684	7173	364594	0.8962	0.1266	1475	7823	0.8962	0.1301	1468	6825
[5000.1.7]: C / H / 300	0.2	0.2	0.8715	0.1035	2363	71630	0.8722	0.1319	177	7098	0.8722	0.1317	177	7087
[5000.1.8]: C / H / 900	0.1	0.2	0.9204	0.1584	5871	326990	0.9232	0.2634	1370	32291	0.9232	0.2650	1351	29118
[5000.2.1]: F / L / 300	0.25	0.7	0.7919	0.0708	4782	37381	0.7912	0.0712	685	29707	0.7884	0.0555	649	16368
[5000.2.2]: F / L / 900	0.15	0.1	0.8536	0.1156	7548	87901	0.8428	0.2746	1485	5575	0.8426	0.2903	1478	4050
[5000.2.3]: F / H / 300	0.2	0.3	0.7946	0.0759	2695	22128	0.7946	0.0798	147	9926	0.7946	0.0798	147	9802
[5000.2.4]: F / H / 900	0.15	0.3	0.8354	0.1223	3964	24811	0.8352	0.2295	313	996	0.8352	0.2300	313	929
[5000.2.5]: C / L / 300	0.2	0.4	0.8019	0.0203	6203	268571	0.7862	0.0319	692	13978	0.7861	0.0321	690	13342
[5000.2.6]: C / L / 900	0.15	0.4	0.8365	0.0257	7476	358538	0.8158	0.0569	1038	8850	0.8157	0.0586	1037	7891
[5000.2.7]: C / H / 300	0.2	0.3	0.7985	0.0434	2981	87816	0.7986	0.0748	143	8600	0.7986	0.0748	143	8486
[5000.2.8]: C / H / 900	0.15	0.3	0.8376	0.0623	4045	113066	0.8375	0.1869	297	2259	0.8375	0.1868	297	2216
[5000.3.1]: F / L / 300	0.2	0.4	0.8677	0.1854	5177	56915	0.8676	0.2282	1404	15288	0.8676	0.2376	1389	13763
[5000.3.2]: F / L / 900	0.1	0.6	0.9143	0.2240	8975	209561	0.9136	0.3275	3181	80936	0.9125	0.3331	2792	50375
[5000.3.3]: F / H / 300	0.15	0.2	0.8663	0.1465	3599	233697	0.8675	0.1579	730	139252	0.8687	0.1598	694	62094
[5000.3.4]: F / H / 900	0.1	0.3	0.9269	0.2571	5508	110540	0.9276	0.2904	1459	52057	0.9276	0.2994	1458	49556
[5000.3.5]: C / L / 300	0.2	0.4	0.8547	0.0707	5288	255626	0.8552	0.1145	940	14745	0.8551	0.1182	938	13384
[5000.3.6]: C / L / 900	0.15	0.3	0.9019	0.0859	6879	320222	0.9024	0.1603	1550	7609	0.9024	0.1665	1537	6399
[5000.3.7]: C / H / 300	0.15	0.3	0.8687	0.0997	3440	341182	0.8723	0.1255	574	76928	0.8729	0.1280	553	42888
[5000.3.8]: C / H / 900	0.1	0.4	0.9200	0.1435	5617	321612	0.9231	0.1921	1341	35060	0.9232	0.1997	1337	32251
Configuration averaged for 300 samples			0.8432	0.1002	4080	140944	0.8422	0.1231	644	28511	0.8421	0.1227	634	17684
Configuration averaged for 900 samples			0.8897	0.1497	6320	201858	0.8875	0.2329	1444	24040	0.8874	0.2387	1405	19932

Each network configuration is described by a network ID [nodes.replicate.configuration] specifying the number of *nodes* (100/1,000/5,000), the *replicate* (1/2/3) and the 8 *configurations*. The 8 configurations 1, . . . , 8 correspond to the IDs in Chap. 1 and depend on marker distance (**C**lose N(1,0.1) / **F**ar N(5,1)) / biological variance (**L**ow N(1,0.1) / **H**igh N(1,0.25)) / population size (300/900). *TP*, True positive; *FP*, false positive. See also Fig. 3.3

to be able to assess the overall effects of the two major pruning steps within our approach (G2 → G4: selection of one candidate edge per eQTL; G4 → G5: removal of edges that most likely stem from indirect effects (TRANSWESD); see Fig. 3.1). We will mainly focus on the AUPR measure since this is the most appropriate one for sparse networks.

As a general trend, we observe that the first (eQTL) pruning step leads in all cases to an improvement of the AUPR, particularly pronounced in the case of large population sizes (see also averaged values in Table 3.1). The second (TRANSWESD) pruning step achieves a significant (but compared to the eQTL pruning lower) AUPR improvement when using the larger population size, whereas only a minor or even no effect can be seen for reconstruction based on the small population with 300 individuals. The effects of the two pruning steps are also well reflected by the number of true positive (TP) and false positive (FP) edges in Table 3.1.

As expected, we see that a larger population size always helps to yield a better reconstruction quality (see also Fig. 3.3). Somewhat surprising was the finding that the best (averaged) AUPR value could be found for the G5 graph of medium size networks (1,000 nodes), here we had expected to see this for networks with 100 nodes.

In the following we will discuss the sensitivity of the reconstruction results with respect to the threshold parameters (t^{Qt} and $d_{\min}$) and the impact of marker distance, biological variance, and population sizes by the example of the first 100-nodes network (networks 100.1.1–100.1.8 in Table 3.1). Similar results can be found for the replicates (100.2.x and 100.3.x) and/or networks of larger size (1000.x.x; 5000.x.x). Figure 3.2 shows for configurations 100.1.1–100.1.8 the resulting AUPR and AUROC performances of the reconstructed G5 networks in the two-dimensional space of meaningful threshold parameters. Clearly, as already outlined above, larger population size (900 samples instead of 300) improves the reconstruction quality (compare odd vs. even numbers of network configurations) although, in line with our results in Flassig et al. (2013), the differences are only moderate. We also see that the optimal threshold regions are similar for all 8 networks. However, one can observe that in the case of low sample size (300) the optimal AUROC/AUPR region is more confined. Thus, the method seems to be fairly robust against a variation of thresholds but an appropriate threshold selection strategy is important for small sample sizes. Generally, the genotype–phenotype threshold t^{QT} for edge detection in G1 seems more sensitive and important than the linkage analysis threshold $d_{\min}$ required in preprocessing. Regarding sensitivity of the performance evaluation, AUROC is much less sensitive to the parameters t^{QT} and $d_{\min}$ than AUPR.

Larger marker distance seems beneficial for reconstruction because genotype correlations are then minimized. This can be seen, for instance, when comparing configuration 2 (marker distance N(5, 1)) with 6 (marker distance N(1, 0.1)) in Fig. 3.2. Partially, weak performance due to small marker distance can be compensated by biological variability (configuration 2 vs. 8). However, in the case of small samples and larger marker distance, larger biological variability decreases performance. This is most likely due to a poor signal-to-noise ratio and can be understood as follows. Interactions between genes are derived from target expression variations induced

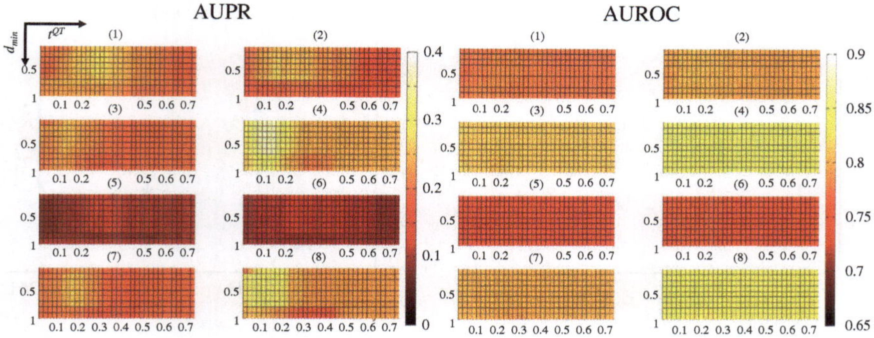

Fig. 3.2 Performance of AUPR (*left*) and AUROC (*right*) of networks 100.1.1–100.1.8 depending on the chosen threshold parameters

by regulator genotype variations. This approach requires sufficient (i) variation of the regulator and (ii) sensitivity of targets with respect to expression variations of the regulator. Variation of the regulator can only be induced by either upstream genes, i.e., the regulator itself is regulated by other genes, and/or by biological variability inducing expression variation in each gene along the sample population. The latter is important for identifying regulator–target interactions of regulators, which have no upstream genes. In this case, the only source of topological informative expression variation is biological variability, which however can only be distinguished from uninformative noise for larger sample sizes.

Figure 3.3 summarizes the AUROC and AUPR performances for all network configurations and sizes averaged over the three network replicates. These results confirm many of the observations made for networks 100.1.x. Again, for our reconstruction algorithm, the worst scenario in terms of AUPR values is the one with small sample size, small marker distance, and small biological variance. We also see that the AUROC is more or less insensitive with respect to sample size and configuration of marker distance/biological variance, but sensitive to the total number of nodes. Specifically, the AUROC is constantly decreased in networks with only 100 nodes compared to 1,000 and 5,000 nodes. This is most likely due to the fact that there are less false negative edges in small compared to large networks (if they have the same connectivity, which is the case for the given dataset) leading to a decreased AUROC. Best network configuration for reconstruction in terms of AUPR values is given by larger samples and large marker distance from which only the first one can be influenced by experimental design. Increased biological variance has noticeable effects on the reconstruction quality for small marker distance. Here, higher biological variance is favorable. The reconstruction quality with respect to network size decreases clearly in one particular case: networks with 5,000 nodes perform poorly in the AUPR values for small sample size (300). Therefore, precision is small in this setting because of too few samples. For 900 samples, precision is raised, resulting into similar AUPR values compared to reconstructions of 100/1,000 node networks.

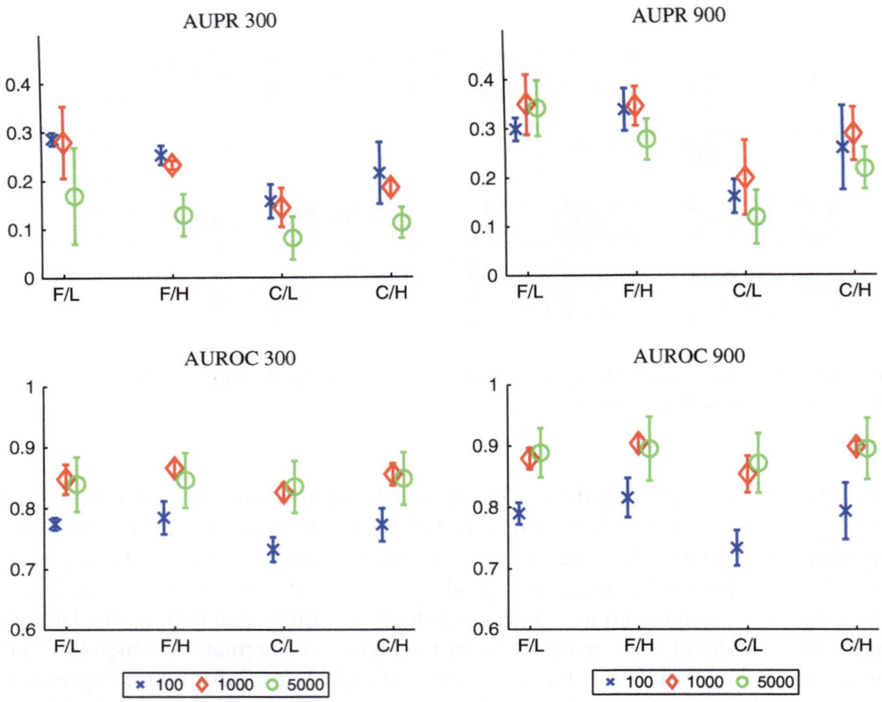

Fig. 3.3 AUPR and AUROC performance averaged over network replicates for different network sizes (100/1,000/5,000 nodes) and samples (300 (*left panel*) or 900 (*right panel*)) grouped according to marker distance (Far/Close) / biological variance (Low/High) configurations

Averaged over all configurations, networks with 1,000 nodes are best reconstructed with respect to AUPR and AUROC values for the eight different configurations.

3.3.2 Prominent Sources of Reconstruction Errors

In the following, we restrict the analysis to (i) a well-identifiable configuration (100.1.4) and (ii) a poorly identifiable configuration (100.1.6). We further restrict our analysis to 900 samples, since the influence of the sample size should be clear from the discussions above. In Fig. 3.4 we show the genotype–phenotype correlation matrix and weight matrix as a density plot. Thereby we have indicated TP (green circles), FP (blue circles), and FN (red circles) in the weight matrix (note that the green and blue circles together describe the reconstructed network G5). In the genotype–phenotype matrix plots we see horizontal gray lines (especially in 100.1.6), which correspond to eQTLs, from which regulators have to be selected, in order to reconstruct the GRN. We see that configuration 100.1.4 tends to have more

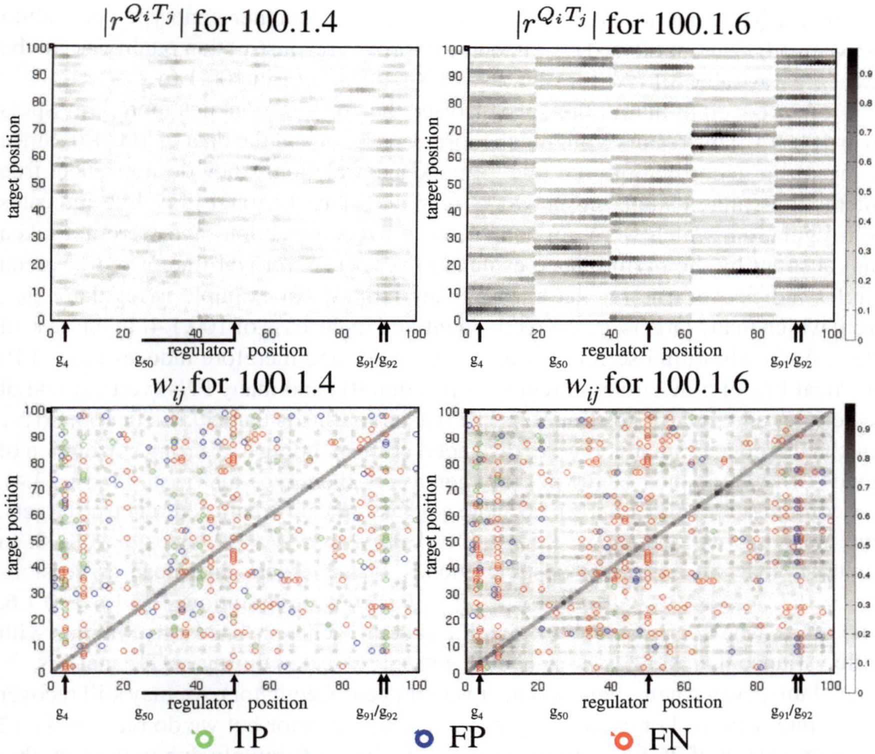

Configuration 100.1.4	Configuration 100.1.6
g_4 : 0 regulators, 26 targets	
μ_E=1.08, σ^2_E=0.19, μ_w=0.11, σ^2_w=0.009, TP=8, FP=7, FN=18	μ_E=0.82, σ^2_E=0.06, μ_w=0.21, σ^2_w=0.022, TP=12, FP=4, FN=14
g_{50} : 1 regulator, 25 targets	
μ_E=0.47, σ^2_E=0.06, μ_w=0.08, σ^2_w=0.007, TP=2, FP=0, FN=23	μ_E=0.48, σ^2_E=0.02, μ_w=0.18, σ^2_w=0.016, TP=0, FP=70 FN=25
g_{91} : 5 regulators, 0 targets	
μ_E=0.35, σ^2_E=0.1, μ_w=0.10, σ^2_w=0.014, TP=0, FP=3, FN=0	μ_E=0.4, σ^2_E=0.03, μ_w=0.19, σ^2_w=0.03, TP=0, FP=14, FN=0
g_{92} : 1 regulator, 21 targets	
μ_E=1.57, σ^2_E=0.43, μ_w=0.11, σ^2_w=0.017, TP=16, FP=5, FN=5	μ_E=1.35, σ^2_E=0.098, μ_w=0.18, σ^2_w=0.02, TP=0, FP=0, FN=21

Fig. 3.4 The *upper panel* shows the genotype–phenotype correlation matrix and the *middle panel* the edge weights (for calculation see Fig. 3.1) of all potential interactions for configurations 100.1.4/100.1.6. *Horizontal gray lines* in the genotype–phenotype correlation matrix correspond to eQTLs, from which regulator genes have to be selected. In the weight matrix, *green* (TP) and *blue circles* (FP) indicate the edges included in the final reconstructed graph G5, whereas *red circles* indicate missed interactions (FN). Some genes (g_4, g_{50}, g_{91}, and g_{92}) were selected for detailed analysis of the TP/FP/FN edges having these genes as regulators (see also Fig. 3.5). Mean expression and its variance of the regulators are given by μ_E and σ_E^2, respectively. Mean weights and weight variances over all target edges of a regulator are indicated by μ_w and σ_w^2

confined eQTLs due to larger marker distances, i.e., smaller genotype correlation between adjacent markers. This of course improves reconstruction quality as can be seen, e.g., in Table 3.1 (AUPR of 0.36 in 100.1.4 vs. 0.12 in 100.1.6).

From the weight matrix plots we also see that 100.1.6 contains more gray spots than 100.1.4. This results from much more correlations in the data of 100.1.6. Since many of these correlations are due to marker correlations, they do not reflect true interactions, thus hampering network inference. The diagonal gray line indicates self-regulation, which were not considered for reconstruction (and were not taken into account by the performance evaluation script). A vertical line of red or green circles indicates a true regulator with many targets. An example is regulator g_{92}, from which many targets are correctly identified in the case of 100.1.4. In the case of 100.1.6, the algorithm selects g_{91} as the regulator and therefore induces many FPs (vertical line of blue circles at regulator position 91) and many FNs (vertical line of red circles at regulator position 92). The reason for this is that eQTLs in 100.1.6 are much larger due to smaller marker distances, corresponding to a strong correlation of genes g_{91}/g_{92} via their genotypes (see genotype–phenotype matrix plot in Fig. 3.4). For configurations 100.1.4/100.1.6, gene g_{92} has 1 true upstream gene, 21 true targets, and mean expressions $\mu_E = 1.57/\mu_E = 1.35$ with $\sigma^2{}_E = 0.43/\sigma_E^2 = 0.098$. In contrast, gene g_{91} has 5 true upstream nodes, 0 true targets, and mean expressions $\mu_E = 0.35/\mu_E = 0.4$ with $\sigma_E^2 = 0.1/\sigma_E^2 = 0.03$ for configurations 100.1.4/100.1.6. Therefore, when deriving the weights for 100.1.6, gene g_{91} has larger weights with little variance than gene g_{92}, thus being wrongly selected during eQTL analysis.

Notably, even when a gene has no upstream gene (regulator), we may still recover target interactions. For example, gene g_4 has no regulator but we do recover 8 / 12 interactions out of 26 for configuration 100.1.4/100.1.6, simply due to the fact, that the expression of gene g_4 is varying due to higher biological variance resulting into expression variations of the targets (see mean edge weights of G4 targets in the table of Fig. 3.4).

Another example for typical challenges of correctly reconstructing interactions from the provided dataset is gene g_{50}. This gene has mean expressions $\mu_E = 0.47/\mu_E = 0.48$ with $\sigma_E^2 = 0.06/\sigma_E^2 = 0.02$ for configurations 100.1.4/100.1.6, with 1 true upstream gene. As the variation in the expressions of gene g_{50} is small, we cannot get any information on its targets superior to variation by noise. Further, even in cases where a regulator is varying strongly it does not necessarily induce variation in the target (see FN histogram and the table in Fig. 3.5). This can happen in cases where a gene has several regulators or if the kinetics of the target activation is in an insensitive range with respect to changes in the regulator (e.g., due to a very low or very large K_m parameter in a Hill function describing the dependency of the target on its regulator). Both effects result into small sensitivity with respect to regulators, thus hampering again the identification of interactions.

In Fig. 3.5 we show three histograms of mean and variance of the regulators' expressions, classified according to whether the (non-)identified target interactions of the regulator are TPs/FPs/FNs. We use network configuration 100.1.4 with optimal threshold parameters as it belongs to the networks with highest reconstruction quality. As expected, regions in the mean–variance expression plane in Fig. 3.5 where we

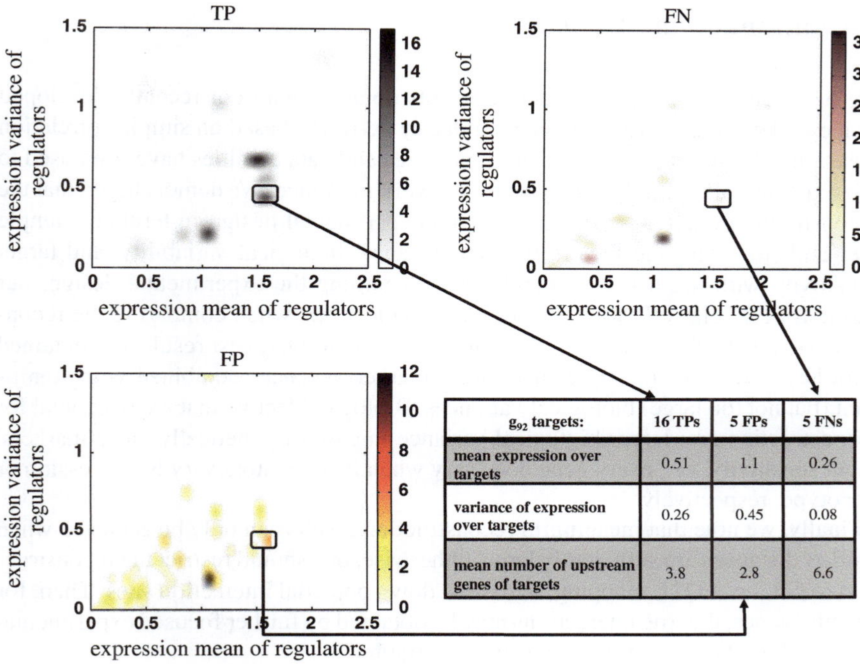

g_{92} targets:	16 TPs	5 FPs	5 FNs
mean expression over targets	0.51	1.1	0.26
variance of expression over targets	0.26	0.45	0.08
mean number of upstream genes of targets	3.8	2.8	6.6

Fig. 3.5 Histograms of expression mean versus expression variance of (non)identified regulators for network configuration 100.1.4 and classified whether the corresponding target regulation is a TP/FP/FN

find TPs also overlap with FP and FN regions. Only for mean and variance levels above 1.2 and 0.4, respectively, FNs and partially FPs are reduced. The drop in FNs is due to the fact that interactions are not missed in the high-level region of the mean–variance plane. Almost independent on the expression mean and variance of a regulator, regulators are sometimes wrongly selected from the eQTLs. This explains why FPs are only slightly reduced in the high-level region.

Interactions of regulators with expression values roughly below 0.5 and variance levels below 0.1 are always mis-classified as either FP or FN. Looking at the mean and variance of the expression levels of the target genes that belong to TP/FP/FN of regulator g_{92} (see table in Fig. 3.4), we see that sufficient variation at a sufficient expression level of the regulator does not guarantee correct identification of (no) interactions. The expression level of the target and its variance also determine classification results. The more inputs a target has, the more likely it is to get an FN since its sensitivity to variation of a specific input node is decreased (see mean expression variance over the FN target genes). False positives are also generated, when the FP targets vary too strongly. In the example of Fig. 3.5, this is probably due to strong biological variance and experimental noise, inducing variations in the FP targets; all five FP targets have a relatively low mean input number of 2.8.

3.4 Summary and Conclusions

We have analyzed the reconstruction results obtained with our recently developed framework for reconstructing gene regulatory networks based on simple correlation measures. Several different network topologies and data qualities have been used to illustrate limitations and challenges for network inference. We demonstrated that the reconstruction quality is influenced by (i) experimental design in terms of sample size and (ii) biological factors (marker distance, biological variability, and target sensitivity with respect to its regulators). Regarding the experimental design, our framework is relatively tolerant to small sample sizes, when comparing the reconstruction results from 300 and 900 sample data. However, best results are obtained with large number of samples and larger marker distances combined with significant (but not too large) biological variances. Biological factors that are beneficial for reconstruction are: larger biological variance in case of genetically close markers, input sensitivity, i.e., every gene does vary when its regulators vary in expression or genotype, respectively.

Finally, we note that meaningful reconstruction results can only be achieved when marker distances are sufficiently large. Otherwise, one should restrict the reconstruction to G3, i.e., eQTL mapping, to narrow down potential interaction sites. Then, for specific genes, the true interactions may be obtained by further focused experimental analysis based on the initial reconstructed graph G3.

References

Bing N, Hoeschele I (2005) Genetical genomic analysis of a yeast segregant population for transcription network inference. Genetics 170:533–542

Flassig RJ, Heise S, Sundmacher K, Klamt S (2013) An effective framework for reconstructing gene regulatory networks from genetical genomics data. Bioinformatics 29(2):246–254

Jansen R, Nap N (2001) Genetical genomics: the added value from segregation. Trends Genet 17:388–391

Jansen R (2003) Studying complex biological systems using multifactorial perturbation. Nat Rev Genet 4:145–151

Keurentjes JJB, Fu J, Terpstra IR et al (2007) Regulatory network construction in Arabidopsis by using genome-wide gene expression quantitative trait loci. Proc Natl Acad Sci USA 104:1708–1713

Klamt S, Flassig RJ, Sundmacher K (2010) TRANSWESD: inferring cellular networks with transitive reduction. Bioinformatics 26:2160–2168

Li H, Lu L, Manly KF et al (2005) Inferring gene transcriptional modulatory relations: a genetical genomics approach. Hum Mol Genet 14:1119–1125

Liu B, de la Fuente A, Hoeschele I (2008) Gene network inference via structural equation modeling in genetical genomics experiments. Genetics 178:1763–1776

Liu B, Hoeschele I, de la Fuente A (2010) Inferring gene regulatory networks from genetical genomics data. In: Das S, Caragea D, Hsu WH, Welch SM (eds) Computational methodologies in gene regulatory networks. IGI Global, Hershey, pp 79–107

Michaelson JJ, Loguercio S, Beyer A (2009) Detection and interpretation of expression quantitative trait loci (eQTL). Methods 48:265–276

Rockman MV, Kruglyak L (2006) Genetics of global gene expression. Nat Rev Genet 7:862–872

Rockman MV (2008) Reverse engineering the genotype-phenotype map with natural genetic variation. Nature 456:738–744

Zhu J, Wiener MC, Zhang C et al (2007) Increasing the power to detect causal associations by combining genotypic and expression data in segregating populations. PLoS Comput Biol 3:e69

Chapter 4
Differential Equation Based Reverse-Engineering Algorithms: Pros and Cons

Gennaro Gambardella, Roberto Pagliarini, Francesco Gregoretti, Gennaro Oliva and Diego di Bernardo

Abstract Ordinary Differential Equations (ODEs) represent a deterministic approach to model gene regulatory networks. ODEs can be used to model changes in gene transcription induced by an external perturbation, such as gene overexpression/downregulation or treatment with a drug. Reverse-engineering algorithms based on ODEs require a choice of a functional form describing the effect of a regulator on its target genes. Here, we focused on an ODE-based reverse engineering algorithm named Network Identification by multiple Regression (NIR) which is rooted on the hypothesis that the regulation exerted by one gene (i.e., a TF) on a target gene can be approximated by a linear function, i.e., the transcription rate of the target gene is proportional to the amount of TF. NIR uses steady-state gene expression measurements and requires knowledge of the genes perturbed in each experiment. We showed that even if originally NIR was created for a different purpose, it can be successfully used to infer gene regulation from an integrated genotype and phenotype dataset. Our results provide evidence of the feasibility of applying reverse-engineering algorithms, such as NIR, to infer gene regulatory networks by integrated analysis of genotype and phenotype.

G. Gambardella (✉) · R. Pagliarini (✉) · D. di Bernardo
Telethon Institute of Genetics and Medicine, Via P. Castellino 111, Naples, Italy
e-mail: gambardella@tigem.it

R. Pagliarini
e-mail: r.pagliarini@tigem.it

D. di Bernardo
e-mail: dibernardo@tigem.it

F. Gregoretti · G. Oliva
Institute of High Performance Computing and Networking, Via P. Castellino 111, Naples, Italy
e-mail: francesco.gregoretti@na.icar.cnr.it

G. Oliva
e-mail: gennaro.oliva@na.icar.cnr.it

A. de la Fuente (ed.), *Gene Network Inference*,
DOI: 10.1007/978-3-642-45161-4_4, © Springer-Verlag Berlin Heidelberg 2013

Keywords ODE · Reverse engineering · Gene networks

4.1 Introduction

The aim of Systems Biology is the elucidation of regulatory and signalling pathways by integrating different levels of experimental evidences in order to build computational models of these complex processes. A paradigm shift is occurring in Biology, which is moving away from being a descriptive science toward becoming a quantitative one, thanks to innovative technologies able to generate vast amounts of quantitative experimental data in a relatively short time (Faith et al. 2007).

Statistical methods for the analysis and visualization of data, such as clustering algorithms and principal component analysis (Andreopoulos et al. 2009; Clarke et al. 2008), have provided a significant contribution to biology. These methods, however, mostly provide information on statistical associations between thousands of RNAs and proteins operating in a cell, but are not able to infer direct causal interactions. Indeed, genes interact with each other in complex regulatory networks which enable the processing of information and the metabolism of nutrients. The identification of these networks represents a fundamental step to understand cellular functions, but it is challenging because of experimental limitations: (i) the number of large-scale measurements available, for example gene expression profiles, is much less than the number of genes, which leads to mathematically ill-posed problems; (ii) the high-level of biological noise, which requires replicated experiments thus increasing experimental costs; (iii): the need of prior knowledge in order to interpret the inferred network in a biologically meaningful way.

Reverse-engineering can be defined as the process of identifying gene regulatory interactions from large-scale experimental data through computational methods. Typically, the aim is to infer (causal) regulatory interactions among genes from gene expression profiles (GEPs). These interactions may not necessarily represent physical ones, but could refer also to indirect regulations via proteins, metabolites, and ncRNA which have not been measured directly. As a result, the meaning of interactions is not well-defined and depends on the formalism selected to model the network. Nevertheless, the inferred networks have several practical uses (Bansal et al. 2007), such as (i) identification of functional modules, (ii) prediction of the system's behavior, (iii) identification of master regulators controlling the major cell functions.

In the recent years, several computational approaches have been proposed to reverse-engineering gene networks from GEPs. Nevertheless, there is not a single "winning approach" which can be generally applied, but each method is tailored to a specific question and to a given type of experimental data (Marbach et al. 2012).

These approaches can be roughly classified into: (i) *boolean networks*, (ii) *bayesian networks*, (iii) *information-theoretic approaches*, (iv) *relevance networks*, (v) *graphical models*, (vi) *neural networks*, (vii) *generating automata* and (viii) *differential equations*. All of these models have advantages and drawbacks, and we refer the interested reader to (Chen 2004; de la Fuente et al. 2004; Kauffman 1969, 1993; Sima et al. 2009; Somogyi and Sniegoski 1996; Szallasi and Liang 1998; Basso et al. 2005) for furthers details.

In this chapter, we will describe the last class of methods, namely Ordinary Differential Equations (ODEs). ODEs represent a deterministic approach to model gene regulatory networks. ODEs relate changes in gene transcription to each other and to an external perturbation, such as overexpression/downregulation of specific genes or the treatment with a drug.

Reverse-engineering algorithms based on ODEs require the choice of a functional form describing the effect of a transcription factor (TF) on its target genes. Nonlinear functions can better capture the regulatory relationship between a TF and a target gene, but these can lead to an exponential rise in the unknown parameters to be estimated.

In what follows, we will focus on an ODE-based reverse-engineering algorithm named *Network Identification by multiple Regression* (*NIR*) (Gardner et al. 2003) which is rooted on the hypothesis that the regulation exerted by one gene (i.e., a TF) on a target gene can be approximated by a linear function, i.e., the transcription rate of the target gene is proportional to the amount of TF.

NIR uses steady-state gene expression measurements and does not require any prior information about gene function or network structure, but only the knowledge of which genes are directly perturbed (overexpressed or downregulated) in each experiment. NIR is based on multiple linear regression models and assumes that only a small subset of genes has been perturbed in each experiment (where one experiment usually corresponds to one measured GEP). It infers sparse gene networks with high accuracy (Cantone et al. 2009).

The chapter is organized as follows. Section 4.2 briefly recalls the ODE model and describes the NIR algorithm. Section 4.3 reports the performances achieved by NIR on the StatSeq benchmark dataset. Finally, further remarks and conclusions are the topic of Sect. 4.4.

4.2 Methods

4.2.1 Model

Let us consider a set $X = \{x_1, x_2, \ldots, x_n\}$ of n genes, where x_i indicates the gene expression measurement of gene i. Let D be the set of m GEPs measuring all the genes at steady-state following m different perturbations (i.e., gene overexpression or knockdown).

A set of ODEs, one for each of the elements of X, can be used to describe gene regulation as a function of other transcripts (de Jong 2002):

$$\dot{x}_i(t) = f_i(x_1, x_2, \ldots, x_n, u, \theta_i), \quad i = 1, 2, \ldots n \tag{4.1}$$

where $\dot{x}_i(t) = \frac{dx_i}{dt}$ is the rate of transcription of transcript i, u is an external perturbation to the system, and θ_i is a set of (unknown) parameters describing causal interaction among genes. f_i can be expanded in a Taylor series around the point in which measurement is made, x_0. If it is assumed that perturbations around this point are small,[1] then it is possible to truncate the series after the first order term in order to obtain a linear approximation of (4.1), that is:

$$\dot{x}_i(t) = \sum_{j=1}^{n} a_{ij}x_j(t) + b_i u_i(t), \quad i = 1, 2, \ldots n \tag{4.2}$$

where $x_i(t)$ and $\dot{x}_i(t)$ are, respectively, the concentration and the rate of change of transcript i measured at time t in one experiment, a_{ij} models the influence of gene x_j on gene x_i, b_i represents the effect of perturbation on x_i, while $u_i(t)$ is the external perturbation to gene x_i at time t (a_{ij} and b_i form the set θ_i of parameters in Eq. (4.1)). In particular, a positive sign for a_{ij} indicates activation, a negative one represents repression, while a zero value means no interaction between x_i and x_j.

In the case of steady-state data, $\dot{x}_i(t) = 0$ for $i = 1, 2, \ldots n$, and then Eq. (4.2) becomes independent of time. Therefore, it can be simplified in the following form:

$$\sum_{j=1}^{n} a_{ij}x_j = -b_i u_i \tag{4.3}$$

for each gene x_i. Since we have m experiments, we can write for gene x_i:

$$[a_{i1}, a_{i2}, \ldots a_{in}] \begin{bmatrix} x_1^1 & x_1^2 & \ldots & x_1^m \\ x_2^1 & x_2^2 & \ldots & x_2^m \\ \vdots & \vdots & \ddots & \vdots \\ x_n^1 & x_n^2 & \ldots & x_n^m \end{bmatrix} = -b_i \left[u_i^1, u_i^2, \ldots u_i^m\right]. \tag{4.4}$$

Equation (4.4) can be written as:

$$\underline{a_i}^T \mathbb{X} = -b_i \underline{u_i}^T \tag{4.5}$$

where $\underline{a_i}$ is a vector whose components are a_{ij}, $j = 1, 2, \ldots n$, $\underline{u_i}$ is a vector whose components are u_i^j, $j = 1, 2, \ldots m$, and $\mathbb{X} \in \mathbb{R}^{n \times m}$. Equation (4.5) can now be solved to obtain the unknowns a_{ij}, if $m \geq n$, whereas the coefficients b_i are usually assumed equals to 1 without loss of generality.

[1] A perturbation u_l is defined small if the system returns to the original steady-state point after removal of u_l and if the magnitude of the response is roughly proportional to the magnitude of u_l (typically the 10 % of the original mRNA concentration).

4.2.2 Algorithm

The NIR algorithm identifies quantitative regulatory relationships among genes, i.e., the parameters a_{ij}, starting from the measured GEPs and the knowledge of which gene(s) have been perturbed in each experiment. In order to apply this approach, the following steps must be performed: (i) controlled perturbations, such as a set of gene overexpression, are delivered to a cell population, (ii) the measured gene expression profiles are collected, and finally (iii) the NIR algorithm is applied to the data in order to learn an ODEs model of the gene network.

The algorithm employs multiple linear regression (Draper and Smith 1998) to estimate the unknown model parameters a_{ij} from Eq. (4.5). NIR requires, as input, the GEPs following each of the m perturbation experiments ($\mathbb{X}$), the knowledge of which genes have been directly perturbed in each perturbation experiment ($\mathbb{U} = \{\underline{u_i}^T\}$, $i = 1 \dots n$), and, optionally, the standard deviation of replicate measurements.

To solve the set of equations in (4.5), a linear regression approach is applied, which identifies the set of unknowns a_{ij} which minimizes the least-squared error between the model prediction and the experimental data. These are the values for which the first derivative of the residual sum of square function is zero under the assumption that the n genes are linearly independent, and can be computed if and only if $m \geq n$. This means that, in order to identify the network model, we have to make at least n distinct perturbations to the considered genes. However, for large sets of genes, it could be impractical to perform a perturbation for each gene and thus the problem would remain undetermined.

To overcome this problem, NIR adopts the sparsity assumption (Newma 2003) which imposes an upper bound, k, on the number of ingoing edges for each gene. That is, it is assumed that each gene can be regulated by at most k other genes, and this upper bound can be defined by the user.

In this way, an undetermined problem is transformed into an overdetermined problem (if we chose $k < m$) which is robust both to measurement noise and incomplete datasets.

In order to infer the best subset of k regulators of each gene, NIR employs the residual sum of squared error minimization criterion. The algorithm therefore has to test, for each gene x_i, all the possible $n!/(k!(n-k)!)$ combinations of k genes. This exhaustive approach is not feasible for a set of genes having cardinality greater than 100, thus NIR employs a heuristic approach developed in (Someren et al. 2001) to reduce the number of sets to test. This approach first computes all the possible solutions with one regulator for each gene, then select the d ones having the smallest least squared error. After that, it computes all the possible solutions with two regulators only for the d intermediate solutions. Then, it again selects the best d solutions, and so on until the number of regulators for each gene is k.

The final output of NIR is the matrix $\mathbb{A}$ which encodes the directed graph, two matrices containing the variance and covariance of each $a_{ij} \in \mathbb{A}$, respectively, and the goodness of fit of the regression.

A drawback of the original NIR was the impossibility to run it on a dataset with more than 100 genes because it would have taken too long to compute (Bansal et al. 2007). However, NIR can be parallelized to handle a larger number of genes in a computationally efficient manner by distributing the overall computation burden among different processors to reduce the total execution time. A parallel version of NIR has been developed in (Gregoretti et al. 2010) and extended in this work. The parallelization is based on the fact that the computational steps of NIR can be performed independently for each gene, and then the parallel implementation assigns different genes to different computing processes. In this way, the parallel implementation scales well as the number of processors increases, has a linear speedup, and the time complexity of the original algorithm can be reduced by one order of magnitude.

4.3 Results

In this section, we reported the results obtained by evaluating the performance of NIR on the StatSeq compendium presented in Chap. 1. Briefly, StatSeq consists of 72 datasets originated from 9 different *"in silico"* gene networks, each simulated under 8 different parameter settings such as population sizes, marker distances, and heritability. For each of the 72 datasets there are two matrices: (i) the gene expression matrix and (ii) the genotype matrix which represents the mutated genes.

To evaluate the NIR performances, we computed two scoring metrics commonly applied to test the efficiency of a binary classifier as a function of the classification threshold: (i) the Area Under the Precision–Recall Curve (AUPR) which summarizes the Precision-Recall tradeoff and (ii) the Area Under the Receiver operating characteristic curve (AUROC) which summarizes the tradeoff between the true positive prediction ratio and the false positive prediction.

Precision is defined as the fraction of retrieved connections that are correctly identified, that is:

$$\text{Precision} = \frac{TP}{(TP + FP)} \tag{4.6}$$

where TP are the true positives and FP are the false positives. Recall, instead, is the fraction of true connections that are retrieved by the algorithm:

$$\text{Recall} = \frac{TP}{(TP + FN)} \tag{4.7}$$

where FN are the false negatives. Hence, while a Precision–Recall curve represents the Precision against the Recall at different thresholds setting, an ROC curve plots the True Positive Rate (TPR) also known as Recall (or Sensitivity) against the False Positive Rate (FPR) at various threshold settings. FPR is also defined as 1-$Specificity$, where Specificity is computed as:

Table 4.1 NIR results for datasets 25–48

Dataset	Network	AUROC	AUPR	$AUPR_{rand}$	$AUPR_{10\,\%}$	$AUPR_{rand10\,\%}$
25	4	54.08	0.96	0.31	0.60	0.03
26	4	55.54	1.43	0.31	0.98	0.03
27	4	51.70	0.49	0.31	0.20	0.03
28	4	54.11	0.90	0.31	0.55	0.03
29	4	51.93	0.54	0.31	0.24	0.03
30	4	52.86	0.62	0.31	0.31	0.03
31	4	50.85	0.41	0.31	0.12	0.03
32	4	51.96	0.50	0.31	0.20	0.03
33	5	56.13	1.47	0.29	0.99	0.03
34	5	57.68	1.86	0.29	1.19	0.03
35	5	53.23	0.61	0.29	0.31	0.03
36	5	55.81	1.20	0.29	0.76	0.03
37	5	53.37	0.57	0.29	0.27	0.03
38	5	54.25	0.72	0.29	0.39	0.03
39	5	51.85	0.47	0.29	0.19	0.03
40	5	53.68	0.61	0.29	0.30	0.03
41	6	55.08	0.10	0.30	0.62	0.03
42	6	56.93	1.65	0.30	1.06	0.03
43	6	53.35	0.59	0.30	0.28	0.03
44	6	54.66	0.87	0.30	0.51	0.03
45	6	53.25	0.58	0.30	0.27	0.03
46	6	54.53	0.72	0.30	0.37	0.03
47	6	52.26	0.41	0.30	0.12	0.03
48	6	53.48	0.54	0.30	0.83	0.03

Results for datasets 1000 genes. AUROC is the area under the entire ROC curve; AUPR the area under the entire precision recall curve; $AUPR_{rand}$ the area under the precision recall curve considering the performance of a random algorithm; $AUPR_{10\,\%}$ the area under the precision recall curve at 10 % of recall; $AUPR_{rand}$10 % the area under the precision recall curve at 10 % of recall considering the performance of a random algorithm

$$\text{Specificity} = \frac{FP}{(FP + TN)} \tag{4.8}$$

where TN are the true negatives.

As we discussed in the previous section, NIR requires two matrices in input: (i) the gene expression profile matrix $\mathbb{X}$ and (ii) the perturbation matrix $\mathbb{U}$. This because NIR absolutely requires the knowledge of which genes have been directly perturbed in each experiment. Here, we used the genotype matrix as the perturbation matrix, with the *"naive"* assumption that a mutated gene will change its expression level. Although NIR is able to infer a signed directed graph, to facilitate comparison among algorithms, we computed the Precision–Recall curve and the ROC curve considering the inferred networks as an unsigned directed graph.

Table 4.2 NIR results for datasets 49–72

Dataset	Network	AUROC	AUPR	$AUPR_{rand}$	$AUPR_{10\,\%}$	$AUPR_{rand10\,\%}$
49	7	53.3606	0.3638	0.059	0.3024	0.006
50	7	55.0744	0.6059	0.059	0.5154	0.006
51	7	51.5462	0.1199	0.059	0.064	0.006
52	7	53.6633	0.3337	0.059	0.271	0.006
53	7	51.9643	0.1256	0.059	0.0688	0.006
54	7	52.8732	0.1883	0.059	0.1288	0.006
55	7	51.1194	0.0863	0.059	0.0313	0.006
56	7	52.0822	0.1256	0.059	0.0685	0.006
57	8	52.8597	0.2571	0.063	0.194	0.006
58	8	54.5249	0.4978	0.063	0.4244	0.006
59	8	51.2068	0.1005	0.063	0.042	0.006
60	8	52.9636	0.2858	0.063	0.2221	0.006
61	8	51.4912	0.1069	0.063	0.0477	0.006
62	8	52.269	0.1449	0.063	0.0836	0.006
63	8	50.8817	0.0833	0.063	0.0255	0.006
64	8	51.7043	0.1109	0.063	0.0512	0.006
65	9	52.5285	0.2525	0.061	0.191	0.006
66	9	54.2944	0.454	0.061	0.3823	0.006
67	9	51.2804	0.1106	0.061	0.0529	0.006
68	9	53.1384	0.2741	0.061	0.2105	0.006
69	9	51.5462	0.1106	0.061	0.0522	0.006
70	9	52.3102	0.1526	0.061	0.0921	0.006
71	9	50.8189	0.0814	0.061	0.0247	0.006
72	9	51.6396	0.1236	0.061	0.0649	0.006

Results for datasets of 5000 genes. AUROC is the area under the entire ROC curve; AUPR the area under the entire precision recall curve; $AUPR_{rand}$ the area under the precision recall curve considering the performance of a random algorithm; $AUPR_{10\,\%}$ the area under the precision recall curve at 10 % of recall; $AUPR_{rand}10\,\%$ the area under the precision recall curve at 10 % of recall considering the performance of a random algorithm

Results from the application of NIR to the StatSeq dataset are reported in Table 4.1 for gene networks of size 1000 and in Table 4.2 for gene networks of size 5000. The label "AUPRrand" refers to the expected performance of an algorithm that randomly connects genes assuming a uniform probability distribution over the set of possible edges. In order to compute this value, we considered a hyper-geometrically distributed random variable whose distribution function and expected value are, respectively:

$$P_x = \frac{\binom{M}{x}\binom{N-M}{n_p-x}}{\binom{N}{x}}, \quad E[X] = M\frac{\binom{N-1}{n_p-1}}{\binom{N}{n_p}} = M\frac{n_p}{N} \tag{4.9}$$

where N is the number of possible edges in the network, M is the number of true edges, and n_p is the number of predicted edges. We then computed the random Precision as follows:

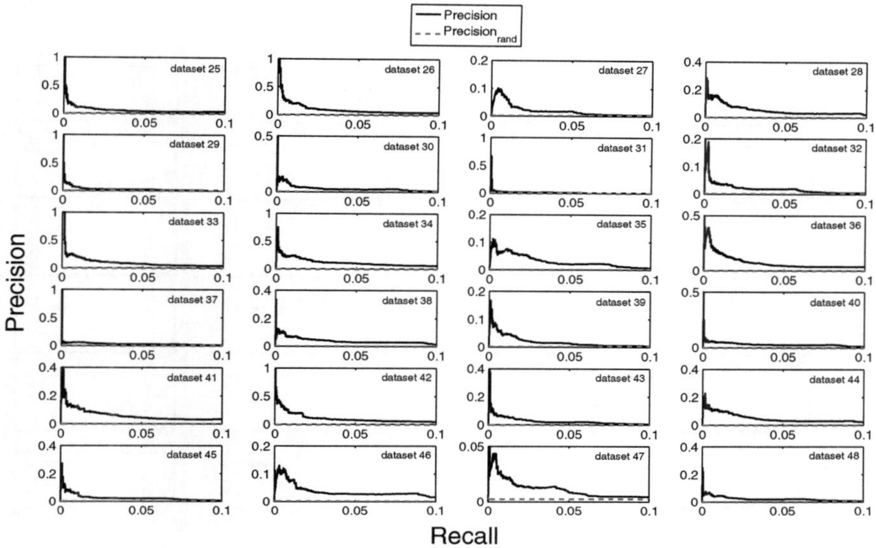

Fig. 4.1 Precision–Recall curve at 10 % of Recall for the 24 networks of size 1000. The Precision (TP/(TP + FP)) versus Recall (TP/(TP + FN)) curve at 10 % of Recall across the 24 datasets composed of 1,000 genes. The dashed line represent the precision of a random algorithm

$$\text{Precision}_{\text{rand}} = \frac{TP_{\text{rand}}}{TP + FP} = \frac{E[X]}{n_p} = \frac{M}{N}. \tag{4.10}$$

From Eq. (4.10), since the width of the Recall interval is always 1, it is possible to obtain the area under the precision–recall curve of a random algorithm simply as the area of a rectangle of base equal to 1 and height equal to $\text{Precision}_{\text{rand}}$:

$$\text{AUPR}_{\text{rand}} = 1 * \text{Precision}_{\text{rand}} \tag{4.11}$$

Since NIR assumes that each gene can be regulated by at most k regulators and $k << n$ (i.e. in our settings $k = 10$ and $n = 1000$ or $n = 5000$), then the algorithm produces an extremely sparse matrix $\mathbb{A}$. Therefore the AUC of the entire Precision–Recall curve could not be considered as a right measure to assess the performance of NIR, because the majority of inferred connections in the network are zero and then randomly sorted. To avoid this bias, we reported in Tables 4.1 and 4.2 also the AUC at a Recall level of 10 % and its comparison with the random (all the values are multiplied by 100 to obtain a percentage value).

At the end of this section, we would like to point out that, since we approximated the perturbation matrix $\mathbb{U}$ with the genotype matrix, the used data were not an ideal input for NIR. However, Tables 4.1, 4.2, Figs. 4.1 and 4.2 show that the results are quite remarkable. Indeed, it is possible to see that NIR performed better than random.

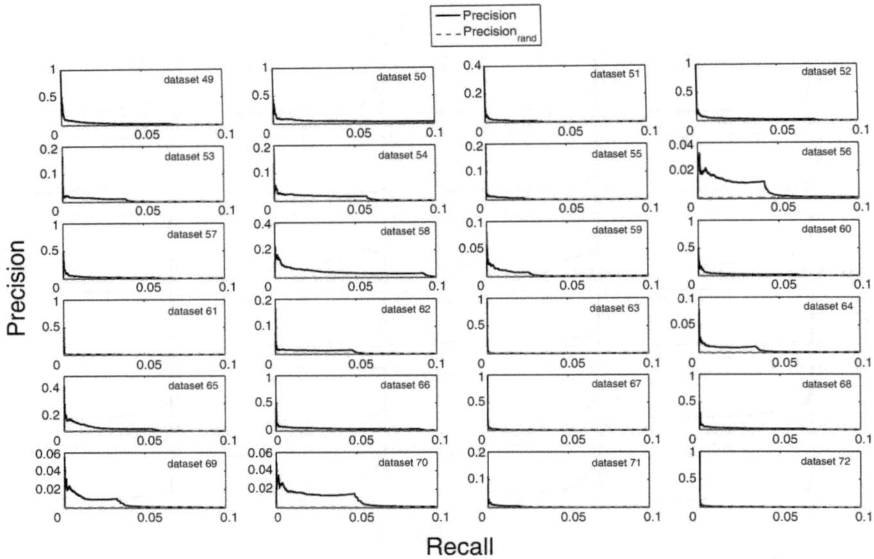

Fig. 4.2 Precision–Recall curve at 10 % of Recall for the 24 networks of size 5000. The Precision (TP/(TP + FP)) versus Recall (TP/(TP + FN)) curve at 10 % of Recall across the 24 datasets composed of 5,000 genes. The dashed line represents the precision of the random algorithm

In addition, for smaller network (i.e., size 100), the performance of NIR increases considerably (i.e., on average AUPR = 4.4 % and AUROC = 56.6 %).

As a further test, we compared NIR with ARACNe (Adam et al. 2006), an inference algorithm based on an information-theoretic approach. This algorithm uses a generalization of Pearson's correlation coefficient called mutual information (*MI*), to measure the degree of independence between two genes. For each pair of genes (x_i, x_j), their MI_{ij} is computed and an edge is added between the two genes depending on a significant threshold to which MI_{ij} is compared. If two genes x_i and x_j are statistically independent then $MI_{ij} = 0$, while a higher MI_{ij} indicates that the two genes are non-randomly associated to each other. However, in experimental setting, an estimated mutual information never equals zero. Under this scenario the recovered gene network would be full connected.

Figure 4.3 shows the comparison between the two algorithms by plotting the Precision–Recall curves at Recall level of 10 % for six networks of size 1000. It is possible to see that, for these datasets, NIR outperforms ARACNe.

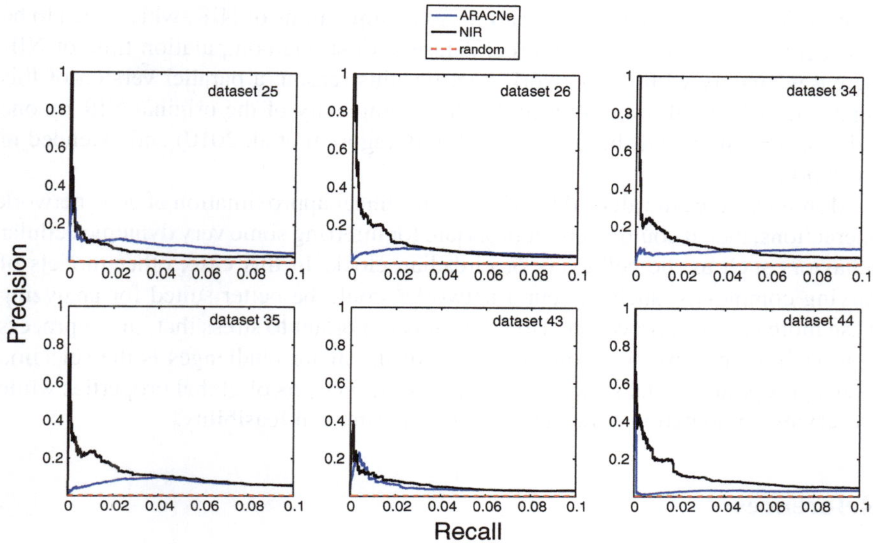

Fig. 4.3 Precision–Recall curve at 10 % of Recall for NIR and ARACNe algorithms. The Precision (TP/(TP + FP)) versus Recall (TP/(TP + FN)) curve at 10 % of Recall for NIR (black line) and ARACNe (blue line) algorithms. Only the first two types of each datasets composed of 1,000 genes have been used. The dashed line represents the precision of the random algorithm

4.4 Discussions and Conclusions

This chapter has been dedicated to NIR, a deterministic reverse-engineering algorithm which quantifies the influence of genes on one another by using measurements of transcptional responses to genetic perturbations and multiple linear regression. NIR uses a set of linear differential equations to describe a gene regulatory network. This approximation of genetic interactions allows NIR to infer a gene network from a small number of GEPs, by limiting the number of regulators for each gene.

We described the application of NIR to steady-state gene expression traits which have been simulated for a population of individuals, based on a complex nonlinear gene network model and a simulated set of genome-wide DNA variants. We would like to point out that the data used in this book are not specific to NIR. Therefore, in order to apply it, we made the *"naive"* assumption that a mutated gene always changes its expression level. This assumption allows us to use the genotype matrix as a perturbation matrix. However, in spite of this assumption, the sign of each perturbation was not known. Despite this limitation, we showed that NIR performs better than the random and better than an approach, such as ARACNe (Adam et al. 2006), which does not use the genotype information. Our results provide evidence of the feasibility of applying reverse-engineerig algorithms, such as NIR, to infer gene regulation by integrated analysis of genotype and phenotype data.

We conclude this chapter by recalling some limitations of NIR, which need to be considered when it is applied on real datasets. First, the computation time of NIR increases quickly with the network size. For this reason, a parallel version of this algorithm, which allows to reduce the time complexity of the original NIR by one order of magnitude, has been developed in (Gregoretti et al. 2010) and extended in this work.

Moreover, since the algorithm is based on a linear approximation of gene network interactions, then it could not be appropriate for inferring some very dynamic cellular behaviors such as the cell-cycle or circadian clock. In this case, other models of varying complexity, such as neural networks, could be better suited for analyzing these more complex processes. However, it is important to stress that, in the process of modeling and reverse-engineering, one of the major challenges is the selection of an appropriate model structure that allows the analysis of global properties while preserving computational tractability, and experimental feasibility.

References

Adam M, Ilya N, Katia B, Chris W, Riccardo F, Andrea C (2006) ARACNE: an algorithm for the reconstruction of gene regulatory networks in a mammalian cellular context. BMC Bioinform 7: S7+

Andreopoulos B, An A, Wang X, Schroeder M (2009) A roadmap of clustering algorithms: finding a match for a biomedical application. Briefings bioinform 10:297–314

Bansal M, Belcastro V, Ambesi-Impiombato A, di Bernardo D (2007) How to infer gene networks from expression profiles. Mol Syst Biol 3:78

Basso K, Margolin AA, Stolovitzky G, Klein U, Dalla-Favera R, Califano A (2005) Reverse engineering of regulatory networks in human B cells. Nat Genet 37:382–390

Cantone I, Marucci L, Iorio F, Ricci MA, Belcastro V, Bansal M, Santini S, di Bernardo M, di Bernardo D, Cosma MP (2009) A yeast synthetic network for in vivo assessment of reverse-engineering and modeling approaches. Cell 137:172–181

Chen PC (2004) A computational model of a class of gene networks with positive and negative controls. Biosystems 73:13–24

Clarke R, Ressom HW, Wang A, Xuan J, Liu MC, Gehan EA, Wang Y (2008) The properties of high-dimensional data spaces: implications for exploring gene and protein expression data. Nat Rev Cancer 8:37–49

de Jong H (2002) Modeling and simulation of genetic regulatory systems: a literature review. J Computl Biol 9:67–103

de la Fuente A, Bing N, Hoeschele I, Mendes P (2004) Discovery of meaningful associations in genomic data using partial correlation coefficients. Bioinformatics 20:3565–3574

Draper NR, Smith H (1998) Applied regression analysis. In: Draper NR, Smith H (eds) Wiley Series in Probability and Statistics, 3rd edn. Wiley, New York

Faith JJ, Hayete B, Thaden JT, Mogno I, Wierzbowski J, Cottarel G, Kasif S, Collins JJ, Gardner TS (2007) Large-scale mapping and validation of escherichia coli transcriptional regulation from a compendium of expression profiles. PLoS Biol 5:e8+

Gardner TS, di Bernardo D, Lorenz D, Collins JJ (2003) Inferring genetic networks and identifying compound mode of action via expression profiling. Sci New York NY 301:102–105

Gregoretti F, Belcastro V, di Bernardo D, Oliva G (2010) A parallel implementation of the network identification by multiple regression (NIR) algorithm to reverse-engineer regulatory gene networks. PLoS ONE 5:e10179

Kauffman SA (1969) Homeostasis and differentiation in random genetic control networks. Nature 224:177–178

Kauffman SA (1993) The origins of order: self-organization and selection in evolution, 1st edn. Oxford University Press, USA

Marbach D, Costello JC, Küffner R, Vega NM, Prill RJ, Camacho DM, Allison KR, The DREAM5 Consortium, Kellis M, Collins JJ, Stolovitzky G (2012) Wisdom of crowds for robust gene network inference. Nat methods 9:796–804

Newman MEJ (2003) The structure and function of complex networks. SIAM REVIEW 45:167–256

Sima C, Hua J, Jung S (2009) Inference of gene regulatory networks using time-series data: a survey. Curr Genomics 10:416–429

Someren EPV, Wessels L, Reinders M, Backer E (2001) Searching for limited connectivity in genetic network models. In: Proceedings of the second international conference on systems biology.

Somogyi R, Sniegoski C (1996) Modeling the complexity of genetic networks: understanding multigenic and pleiotropic regulation. Complexity 1:45–63

Szallasi Z, Liang S (1998) Modeling the normal and neoplastic cell cycle with realistic Boolean genetic networks: Their application for understanding carcinogenesis and assessing therapeutic strategies. Pac Symp Biocomputing 3:66–76

Chapter 5
Gene Regulatory Network Inference from Systems Genetics Data Using Tree-Based Methods

Vân Anh Huynh-Thu, Louis Wehenkel and Pierre Geurts

Abstract One of the pressing open problems of computational systems biology is the elucidation of the topology of gene regulatory networks (GRNs). In an attempt to solve this problem, the idea of systems genetics is to exploit the natural variations that exist between the DNA sequences of related individuals and that can represent the randomized and multifactorial perturbations necessary to recover GRNs. In this chapter, we present new methods, called GENIE3-SG-joint and GENIE3-SG-sep, for the inference of GRNs from systems genetics data. Experiments on the artificial data of the StatSeq benchmark and of the DREAM5 *Systems Genetics* challenge show that exploiting jointly expression and genetic data is very helpful for recovering GRNs, and one of our methods outperforms by a large extent the official best performing method of the DREAM5 challenge.

5.1 Introduction

Networks are commonly used in biological research to represent information. In this chapter, we focus on GRNs. These networks represent regulatory interactions among genes that happen at the level of transcription, through transcription factors. They often offer a simplified view of gene regulation, and are usually represented by graphs where each node corresponds to a gene and an edge directed from one gene to another gene indicates that the first gene codes for a transcription factor that regulates the rate of transcription of the second gene.

V. A. Huynh-Thu · L. Wehenkel · P. Geurts (✉)
Department of EE and CS and GIGA-R , University of Liège, Liège, Belgium
e-mail: P.Geurts@ulg.ac.be

V. A. Huynh-Thu
e-mail: vahuynh@ulg.ac.be

L. Wehenkel
e-mail: L.Wehenkel@ulg.ac.be

A. de la Fuente (ed.), *Gene Network Inference*,
DOI: 10.1007/978-3-642-45161-4_5,

Edges in regulatory networks can be directed or undirected. An undirected edge connecting two genes indicates that there exists a transcriptional regulatory interaction between these two genes, while a directed edge means furthermore that the source gene regulates the expression of the target gene. Edges can also be signed. When a gene is connected to another gene, a positive (resp. negative) sign indicates that the former is an activator (resp. repressor) of the latter. In this chapter, we focus on directed unsigned networks. The targeted networks are thus graphs with p nodes, where an edge directed from one gene i to another gene j indicates that gene i (directly) regulates, either positively or negatively, the expression of gene j $(i, j = 1, \ldots, p)$.

The problem of the inference of regulatory networks has been studied for many years in the literature and many algorithms already exist. The authors De Smet and Marchal (2010) proposed a categorization of these methods. First, they distinguish supervised from unsupervised methods. Supervised methods exploit prior partial knowledge of the network to guide the network inference, while unsupervised methods do not assume any prior knowledge. There are also direct methods, which consider only individual interactions, and module-based methods, which search for sets of genes that are regulated by the same transcription factors. Finally, non-integrative methods only use expression data for the inference, while integrative methods also use other kinds of information besides expression data, e.g. counts of sequence motifs that serve as binding sites for transcription factors.

Among integrative methods, one can also find methods exploiting systems genetics data. The goal of systems genetics is to exploit the natural variations that exist between the DNA sequences of related individuals in a segregating population and that can represent the randomized and multifactorial perturbations necessary to recover (GRNs) (Jansen and Nap 2001; Jansen 2003). In such a study, two strains that are widely separated in terms of genetic background are crossed and their children are self-crossed during several generations in order to produce a recombinant inbred line (RIL) segregating population. The genomes of the individuals of this population comprise random segments of the genomes of the two original parents and genetic differences can therefore be detected between them, representing multifactorial genetic perturbations. Each individual is then analyzed by microarray expression profiling as well as by genetic marker analysis.

Multiple methods have been developed to infer GRNs from systems genetics data. Several methods infer causal regulatory relationships among pairs of genes, including procedures that rely on statistical tests to identify causal links (Chen et al., 2007) and approaches based on the fitting of causal models (Kulp and Jagalur 2006; Schadt et al. 2005). Other methods are based on the analysis of the correlation between expression profiles of genes located in a particular genomic region and expression profiles of genes that are potentially affected by the markers located in this region (Bing and Hoeschele, 2005). Methods that study the regulatory relationships at a systems-level include approaches based on Bayesian networks (Zhu et al. 2007; Li et al. 2005; Vignes et al. 2011), structural equation models (Li et al. 2006; Liu et al. 2008), and the orientation of the edges of an undirected network using genetic markers as causal anchors (Aten et al. 2008; Chaibub Neto et al. 2008). Random Forests have also been successfully used for expression quantitative trait loci (eQTL) mapping (Michaelson et al., 2010).

In this chapter, we propose new methods, based on ensembles of regression trees, for the inference of regulatory networks from systems genetics data. According to the categories of De Smet and Marchal (2010), these methods are direct (we do not search for modules) and unsupervised (we do not assume any prior knowledge of the network).

The chapter is structured as follows. Section 5.2 describes our network inference methods. Section 5.3 shows the results obtained with these methods on the StatSeq compendium as well as on the datasets of the DREAM5 *Systems Genetics* challenge. Finally, Sect. 5.4 concludes the chapter and discusses some ideas for further developments.

5.2 Methods

We assume that we have at our disposal a dataset containing the steady-state expression levels of p genes measured in N individuals, as well as the genotype value of one genetic marker for each of these genes in the same N individuals:

$$LS = \{(\mathbf{e}_1, \mathbf{m}_1), (\mathbf{e}_2, \mathbf{m}_2), \ldots, (\mathbf{e}_N, \mathbf{m}_N)\}, \tag{5.1}$$

where $\mathbf{e_k} \in \mathbb{R}^\mathbf{p}$ and $\mathbf{m_k} \in \{\mathbf{0}, \mathbf{1}\}^\mathbf{p}, \mathbf{k} = \mathbf{1}, \ldots, \mathbf{N}$ are, respectively, the vectors of expression levels and genotype values of the p genes in the kth individual:

$$\begin{cases} \mathbf{e}_k = (e_k^1, e_k^2, \ldots, e_k^p)^\top, \\ \mathbf{m}_k = (m_k^1, m_k^2, \ldots, m_k^p)^\top. \end{cases} \tag{5.2}$$

Note that we suppose that the individuals come from a RIL population and are hence homozygous. Each genetic marker can thus have two possible genotype values only.

From this dataset, our goal is to infer a gene regulatory network, i.e., to make a prediction of the underlying regulatory links between genes. Many network inference algorithms work first by providing a ranking of the potential regulatory links from the most to the less significant. A practical network prediction is then obtained by setting a threshold on this ranking. In this chapter, we focus only on the first task and the question of the choice of an optimal confidence threshold, although important, is left open.

A network inference algorithm is thus defined in this chapter as a procedure that assigns weights $w_{i,j} \geq 0 (i, j = 1, \ldots, p)$ to putative regulatory links from any gene i to any gene j, with the aim of yielding larger values for weights that correspond to actual regulatory interactions.

To infer GRNs from systems genetics data, we propose two extensions of a method called GENIE3 (Huynh-Thu et al., 2010) that exploits tree-based ensemble methods for the inference of networks from expression data. As in the GENIE3 procedure, our two extensions decompose the problem of recovering a network of p genes into p feature selection subproblems, where each of these subproblems consists in identifying the regulators of one of the genes of the network. This idea has also been exploited in other methods, such as MRNET (Meyer et al., 2007), the Graphical Lasso (Meinshausen and Bühlmann, 2006), or the meta-analysis developed by Vignes et al. (2011).

5.2.1 Network Inference as a Feature Selection Problem

To infer GRNs from systems genetics data, the two procedures that we propose make the assumption that the expression of each gene j in a given individual is a function of the expression and genotype values of the other genes of the network in the same individual (plus some random noise). The first procedure, called **GENIE3-SG-joint**, learns a single predictive model from both expression and genetic data, while the second procedure, called **GENIE3-SG-sep**, learns two separate predictive models, one based on the genetic markers and the other based on the expression data. Both methods then compute, for each gene $i \neq j$, two scores $w^e_{i,j}$ and $w^m_{i,j}$, measuring, respectively, the importances of the expression and of the marker of gene i when predicting the expression of gene j. Depending on the method, the computation of $w^e_{i,j}$ and $w^m_{i,j}$ is different. These two scores are then aggregated to obtain a single weight $w_{i,j}$ for the regulatory link directed from gene i to gene j.

We first describe the procedures for training the predictive models and computing the importance scores. We then discuss aggregation techniques, which are common to both approaches, to obtain the final weights.

5.2.1.1 GENIE3-SG-Joint

The GENIE3-SG-joint procedure assumes that a unique model f_j explains the expression of a gene j in a given individual, knowing the expression levels and the genotype values of the different genes of the network:

$$e^j_k = f_j(\mathbf{e}^{-j}_k, \mathbf{m}_k) + \varepsilon_k, \forall k, \tag{5.3}$$

where ε_k is a random noise and $\mathbf{e_k^{-j}}$ is the vector containing the expression levels of all the genes except gene j in the kth individual:

$$\mathbf{e}^{-j}_k = (e^1_k, \ldots, e^{j-1}_k, e^{j+1}_k, \ldots, e^p_k)^\top. \tag{5.4}$$

We further make the assumption that the function f_j only exploits the expression levels in $\mathbf{e_k^{-j}}$ and/or the genotype values in $\mathbf{m_k}$ of the genes that are direct regulators of gene j, i.e., genes that are directly connected to gene j in the targeted network.

Notice that $\mathbf{m_k}$ contains the genotype value of gene j. Indeed, it often happens that a genetic marker contributes to the expression of the gene in which it is located (*cis*-acting polymorphism). Including the marker $\mathbf{m^j}$ of gene j in the input variables thus avoids to wrongly attribute to another regulator the part of the expression of gene j that is actually explained by $\mathbf{m^j}$.

Recovering the regulatory links pointing to gene j thus amounts to finding those genes whose expression and/or genetic marker are predictive of the expression of the target gene. In machine learning terminology, this can be considered as a feature selection problem (in regression) for which many solutions exist (Guyon and Elisseeff 2003; Saeys et al. 2007). We assume here the use of a feature ranking technique that, instead of directly returning a feature subset, yields a ranking of the features from the most relevant to the least relevant for predicting the output.

The GENIE3-SG-joint procedure is illustrated in Fig. 5.1 and works as follows:

- For $j = 1$ to p:
  - Generate the learning sample of input–output pairs for gene j:

$$LS^j = \{((\mathbf{e}_k^{-j}, \mathbf{m}_k), e_k^j), k = 1, \ldots, N\}. \tag{5.5}$$

  - Use a feature ranking technique on LS^j to compute confidence levels $w_{i,j}^e (i \neq j)$ and $w_{i,j}^m$, $i = 1, \ldots, p$, respectively, for the expression and the genetic marker of input gene i.
  - Aggregate $w_{i,j}^e$ and $w_{i,j}^m$ to get a weight $w_{i,j}$ for each gene $i \neq j$ (see Sect. 5.2.1.3).
- Use $w_{i,j}$ as weight for the regulatory link $i \to j$ and get a ranking of all links.

5.2.1.2 GENIE3-SG-Sep

In the second proposed procedure, GENIE3-SG-sep, we assume that two different models f_j^e and f_j^m can both explain the expression of a gene j in a given individual, either from the expression levels of the other genes, or from the genotype values:

$$\begin{cases} e_k^j = f_j^e(\mathbf{e}_k^{-j}) + \varepsilon_k, \forall k, \\ e_k^j = f_j^m(\mathbf{m}_k) + \varepsilon_k', \forall k. \end{cases} \tag{5.6}$$

The functions f_j^e and f_j^m are therefore, respectively, learned from two different learning samples. The method is illustrated in Fig. 5.2 and works as follows:

- For $j = 1$ to p:
  - Generate two learning samples of input–output pairs for gene j:

$$\begin{aligned} LS_e^j &= \{(\mathbf{e}_k^{-j}, e_k^j), k = 1, \ldots, N\}, \\ LS_m^j &= \{(\mathbf{m}_k, e_k^j), k = 1, \ldots, N\}. \end{aligned} \tag{5.7}$$

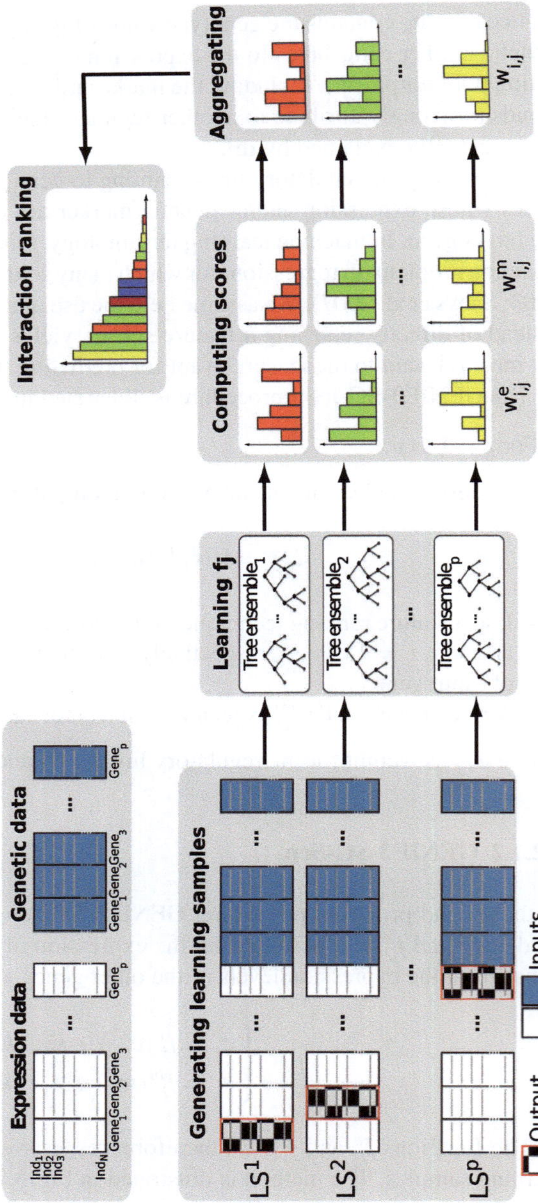

Fig. 5.1 GENIE3-SG-joint procedure. For each gene $j = 1, \ldots, p$, a learning sample LS^j is generated with expression levels of gene j as output values and expression levels and genotypes values of all the other genes as input values. A function f_j is learned from LS^j and confidence levels $w^e_{i,j}$ and $w^m_{i,j}$ are computed for the expression and genotype value of each input gene i respectively. These levels are then aggregated for each input gene and a ranking of all regulatory links is obtained

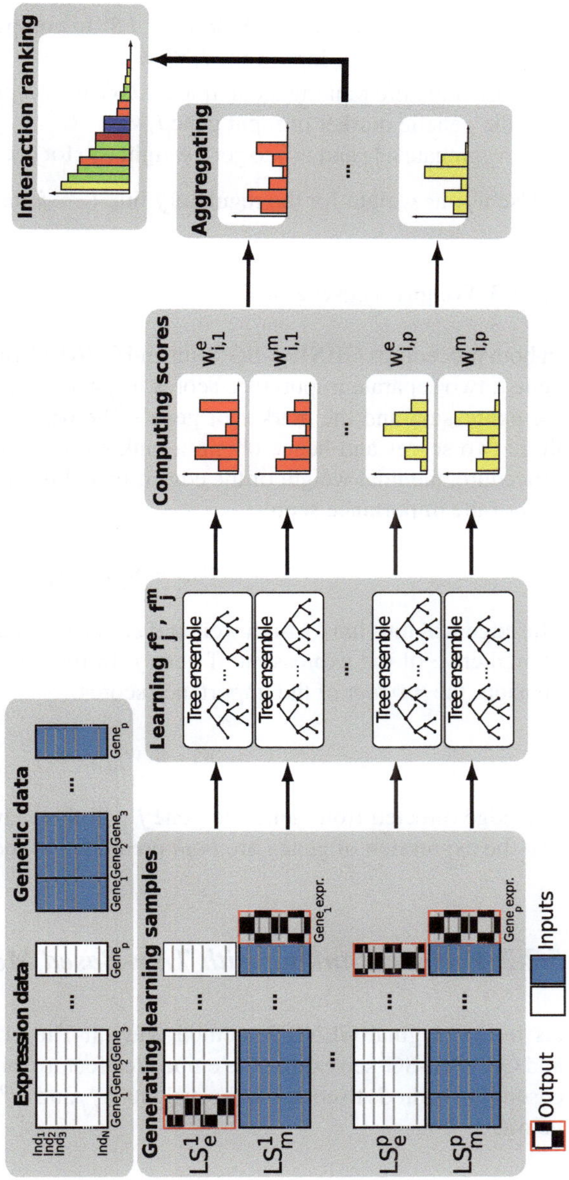

Fig. 5.2 GENIE3-SG-sep procedure. For each gene $j = 1, \ldots, p$, two learning samples LS_e^j and LS_m^j are generated. In both learning samples, the output values are the expression levels of gene j. In LS_e^j the input values are the expression levels of all the other genes, while in LS_m^j the input values are the genotypes values. Functions f_j^e and f_j^m are, respectively, learned from LS_e^j and LS_m^j, and confidence levels $w_{i,j}^e$ and $w_{i,j}^m$ are computed for the expression and genotype value of each input gene i respectively. These levels are then aggregated for each input gene and a ranking of all regulatory links is obtained

  - Use a feature ranking technique on LS_e^j to compute the confidence level $w_{i,j}^e$ of the expression of input gene i, $\forall i \neq j$.
  - Use a feature ranking technique on LS_m^j to compute the confidence level $w_{i,j}^m$ of the genetic marker of input gene i, $\forall i$.
  - Aggregate $w_{i,j}^e$ and $w_{i,j}^m$ to get a weight $w_{i,j}$ for each gene $i \neq j$ (see Sect. 5.2.1.3).
- Use $w_{i,j}$ as weight for the regulatory link $i \rightarrow j$ and get a ranking of all links.

5.2.1.3 Weight Aggregation

In both procedures GENIE3-SG-joint and GENIE3-SG-sep, we obtain for each input gene i, two separate importance scores $w_{i,j}^e$ and $w_{i,j}^m$, corresponding, respectively, to the expression and the marker of gene i. We propose two procedures to aggregate these two scores and hence obtain a ranking of regulatory interactions. In the first procedure, the final weight of the edge directed from gene i to gene j is given by the *sum* of the importance scores:

$$w_{i,j} = w_{i,j}^e + w_{i,j}^m. \tag{5.8}$$

The edge will thus have a high weight if *either* the marker *or* the expression of gene i is predictive of the expression of gene j. In the second aggregation procedure, we consider the *product* of the importance scores:

$$w_{i,j} = w_{i,j}^e \times w_{i,j}^m. \tag{5.9}$$

The edge directed from gene i to gene j will thus have a high weight if the marker *and* the expression of gene i are *both* predictive of the expression of gene j.

5.2.2 *Feature Ranking with Tree-Based Methods*

As in the original GENIE3 method (Huynh-Thu et al., 2010), GENIE3-SG-joint and GENIE3-SG-sep exploit the embedded feature ranking mechanism of tree-based ensemble methods to compute the weights $w_{i,j}^e$ and $w_{i,j}^m$. These methods are described below.

5.2.2.1 Tree-Based Ensemble Methods

Among supervised learning methods, which allow to learn a predictive model or function f_j from observed data, one can find methods based on regression trees (Breiman et al., 1984). The basic idea of regression trees is to recursively split the learning sample with binary tests each based on one input variable, trying to reduce as much as possible the variance of the output variable in the resulting subsets of samples. Candidate splits for numerical variables typically compare the input variable values with a threshold that is determined during the tree growing.

Single trees are usually much improved by ensemble methods, which average the predictions of several trees, such as Bagging (Breiman, 1996) or Random Forests (Breiman, 2001). In a Bagging ensemble, each tree is built from a bootstrap sample of the original learning sample. The Random Forests method adds an extra level of randomization compared to the Bagging; at each test node, K attributes are selected at random among all candidate attributes before determining the best split.

5.2.2.2 Variable Importance Measure

One of the most interesting characteristics of tree-based methods is the possibility to compute from a tree a variable importance measure that allows us to rank the input features according to their relevance for predicting the output. In our experiments, we consider a measure that computes, at each test node $\mathcal{N}$, the total reduction of the variance of the output variable due to the split, defined by Breiman et al. (1984):

$$I(\mathcal{N}) = \#S.\text{Var}(S) - \#S_t.\text{Var}(S_t) - \#S_f.\text{Var}(S_f), \tag{5.10}$$

where S denotes the set of samples that reach node $\mathcal{N}$, S_t (resp. S_f) denotes its subset for which the test is true (resp. false), Var(.) is the variance of the output variable in a subset, and # denotes the cardinality of a set of samples. For a single tree, the overall importance w of one variable is then computed by summing the I values of all tree nodes where this variable is used to split. Those attributes that are not selected at all obtain a zero value of their importance, and those that are selected close to the root node of the tree typically obtain high scores. Attribute importance measures can be easily extended to ensembles, simply by averaging importance scores over all trees in the ensemble. The resulting importance measure is then even more reliable because of the variance reduction effect resulting from this averaging (Hastie et al., 2009).

5.2.2.3 Regulatory Link Ranking

In the GENIE3-SG-joint and GENIE3-SG-sep procedures, the different tree-based models that are generated yield importance scores $w^e_{i,j}$ and $w^m_{i,j}$ for each pair of genes (i, j), computed as sums of variance reductions in the form (5.10). The sum of the importance scores of all input features for a tree is usually very close to the initial total variance of the output. In the case of the GENIE3-SG-joint procedure, we thus have for each target gene j:

$$\sum_{i \neq j}^{p} w^e_{i,j} + \sum_{i=1}^{p} w^m_{i,j} \approx N.\text{Var}_j(LS^{\prime j}), \tag{5.11}$$

where LS'^j is the learning sample from which the tree was built (i.e., a bootstrap sample of LS^j for the Random Forests and Bagging methods) and $\mathrm{Var}_j(LS'^j)$ is the variance of the target gene j estimated in the corresponding learning sample.

Similarly, for the GENIE3-SG-sep procedure, we have:

$$\begin{cases} \sum_{i \neq j} w^e_{i,j} \approx N.\mathrm{Var}_j(LS'^j_e), \\ \sum_i w^m_{i,j} \approx N.\mathrm{Var}_j(LS'^j_m), \end{cases} \tag{5.12}$$

where LS'^j_e and LS'^j_m are the learning samples generated from the expression and genotype data respectively.

As a consequence, if we trivially use the scores $w^e_{i,j}$ and $w^m_{i,j}$ to order the regulatory links, this is likely to introduce a positive bias for regulatory links directed towards the most highly variable genes. To avoid this bias, we first normalize the expression of the target gene j so that it has a unit variance in the training set (LS^j for GENIE3-SG-joint, LS^j_e and LS^j_m for GENIE3-SG-sep), before applying the tree-based ensemble method:

$$\mathbf{e}^j \leftarrow \frac{\mathbf{e}^j}{\sigma^j}, \ \forall j, \tag{5.13}$$

where $\mathbf{e}^j \in \mathbb{R}^N$ is the vector of expression levels of gene j in all N experiments and σ^j denotes its standard deviation. This normalization indeed implies that the different importance scores inferred from different models predicting the different gene expressions are comparable.

5.2.2.4 Computational Complexity

The computational complexity of the Random Forests and Bagging algorithms is $O(TKN \log N)$, where T is the number of trees, N is the dataset size, and K is the number of randomly selected variables at each node of a tree (in the case of Bagging, K is equal to the number of input variables). The complexities of GENIE3-SG-joint and GENIE3-SG-sep are thus of the order of $O(pTKN \log N)$ since these methods require to build, respectively, one and two ensemble(s) of trees for each of the p genes. The complexities are thus log linear with respect to the number of measurements and, at worst, quadratic with respect to the number of genes (when $K = 2p - 1$ for GENIE3-SG-joint and $K = p$ for GENIE3-SG-sep).

To give an idea of the computing times, with our MATLAB®[1] implementations of the methods, GENIE3-SG-sep and GENIE3-SG-joint take, respectively, about 1 and 3 h to infer a network of 1,000 genes from 300 individuals, when K is fixed to the square root of the number of input variables and 1,000 trees are grown in each ensemble. In the worst-case scenario (5,000 genes, 900 individuals, and K equal to the number of input variables), GENIE3-SG-sep and GENIE3-SG-joint would,

[1] http://www.mathworks.com/.

respectively, take 4 months and more than a year to infer the network on a single computer. To reduce computing times, the two algorithms can be trivially parallelized on a computing grid (with one separate computing process for each gene and/or tree).

5.3 Results

After a presentation of the performance metrics, this section presents the results that we obtained when we applied the proposed procedures to two series of synthetic datasets: the StatSeq datasets (Sect. 5.3.2) and the datasets of the DREAM5 *Systems Genetics* challenge (Sect. 5.3.3).

5.3.1 Performance Metrics

Each of our algorithms provides a ranking of the regulatory links from the most confident to the less confident. To evaluate such a ranking independently of the choice of a specific threshold, we used the precision–recall (PR) curve and the area under this curve (AUPR). The PR curve plots, for different thresholds on the weights of the links, the proportion of true positives among all predictions (precision) versus the percentage of true positives among those to be retrieved (recall). A perfect ranking, i.e., a ranking where all the positives are located at the top of the list, yields an AUPR equal to one, while a random ranking results in an AUPR close to the proportion of positives (i.e., close to zero since the proportion of true links among all possible links in a network is usually very low).

5.3.2 Experiments on the StatSeq Datasets

5.3.2.1 Description of the Data

The StatSeq compendium[2] comprises 72 datasets generated from nine different networks. These networks can be divided into three groups of networks of 100, 1,000, and 5,000 genes respectively. For each individual in a dataset, the gene expression levels are provided as well as the genotype value of one genetic marker for each gene. For each of the nine networks, datasets have been generated under eight different setting configurations, by combining different population sizes (300 or 900 individuals), distances between the genetic markers (large or small), and heritability (large or small), as shown in Table 5.1. All networks and datasets were generated using SysGenSIM[3] 1.0.2 (Pinna et al., 2011). The reader can refer to Chap. 1 for details about the StatSeq compendium.

---

[2] http://sysgensim.sourceforge.net/datasets.html.

[3] http://sysgensim.sourceforge.net/.

Table 5.1 Setting configurations for data simulation

Configuration	Marker distance*	Heritability	Population size
1	$\sim\mathcal{N}(5, 1)$	High	300
2	$\sim\mathcal{N}(5, 1)$	High	900
3	$\sim\mathcal{N}(5, 1)$	Low	300
4	$\sim\mathcal{N}(5, 1)$	Low	900
5	$\sim\mathcal{N}(1, 0.1)$	High	300
6	$\sim\mathcal{N}(1, 0.1)$	High	900
7	$\sim\mathcal{N}(1, 0.1)$	Low	300
8	$\sim\mathcal{N}(1, 0.1)$	Low	900

*Means and standard deviations are expressed in centimorgans

5.3.2.2 Comparison of Tree-Based Methods

We applied the GENIE3-SG-joint and GENIE3-SG-sep methods on the StatSeq datasets, using the Random Forests algorithm with the main parameter K fixed to the square root of the number of input variables, as well as the Bagging procedure (equivalent to Random Forests with K fixed to the number of input variables). Ensembles of 1,000 trees were grown in each case, except when we used Bagging to infer networks of 5,000 genes. In that case, only 100 trees were grown in order to reduce the computational burden.[4]

Figure 5.3 shows the AUPR scores obtained with the different combinations. Bagging typically yields better performances than Random Forests, whatever the combination. Lower scores are obtained only with GENIE3-SG-joint (using the product of the weights $w^e_{i,j}$ and $w^m_{i,j}$) for some networks. Therefore, all results shown in the remainder of this chapter will be those obtained with the Bagging procedure.

5.3.2.3 Performance of the GENIE3 Methods

GENIE3-SG-joint versus GENIE3-SG-sep

Figure 5.4 shows the performances of the different GENIE3 procedures. Given an aggregation procedure (either sum or product of the importance scores), better performances are obtained when two separate models are, respectively, learned from the two types of data (GENIE3-SG-sep), instead of one single model (GENIE3-SG-joint). The worse performance of GENIE3-SG-joint can be potentially explained by the fact that when the inputs comprise continuous and discrete variables (with a low number of categories), the Bagging method has a positive bias for the continuous variables when selecting a variable at a test node (Strobl et al., 2007). Indeed, since a continuous variable provides more possible cut-points than a variable with a low number of categories, it has more chance to provide the highest variance reduction

[4] Note that, for the smaller networks, we do not observe significant differences in performance when reducing the number of trees from 1,000 to 100.

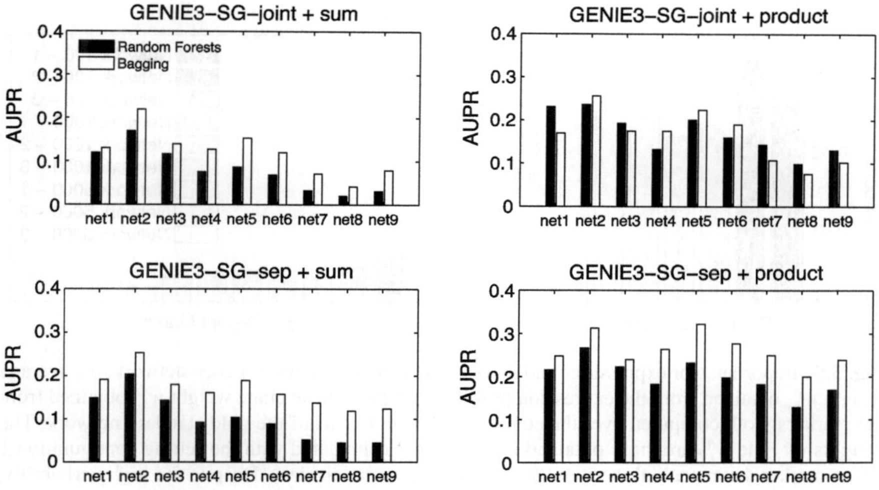

Fig. 5.3 Comparison of tree-based methods. The Bagging method typically yields better performances than Random Forests. The AUPR values of each method were averaged over the eight datasets corresponding to each of the nine networks

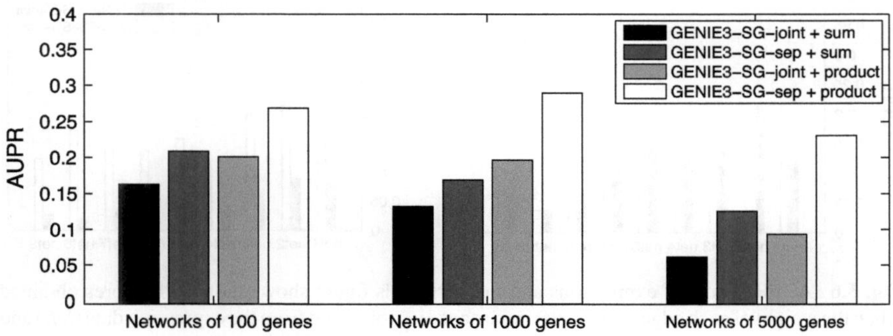

Fig. 5.4 Performances of inference methods. GENIE3-SG-sep yields better performances than GENIE3-SG-joint and higher AUPR scores are obtained by taking the product of the weights $w^e_{i,j}$ and $w^m_{i,j}$ rather than their sum. The AUPR scores of each method were averaged over the 24 datasets related to each network size

on the local node, and hence to be selected for the test, even if it is actually less or equally informative globally. Therefore, in GENIE3-SG-joint, which learns a joint model from the gene expression values (continuous variables) and from the genotype values (discrete variables), the importance $w^m_{i,j}$ of the marker of each gene i is systematically lower than the importance $w^e_{i,j}$ of its expression, as shown in Fig. 5.5. Moreover, the ranking of interactions obtained from the importances $w^m_{i,j}$ is significantly less accurate with GENIE3-SG-joint compared to GENIE3-SG-sep, while the rankings obtained from $w^e_{i,j}$ are equally good (Fig. 5.6).

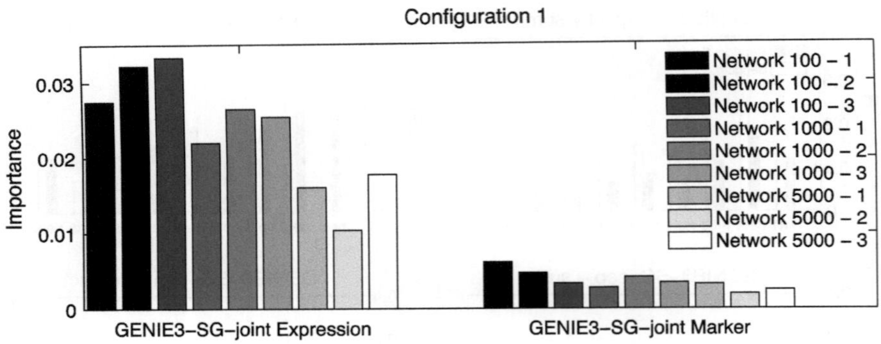

Fig. 5.5 Importance of expressions and markers. This figure shows, for each network, the average weight $w^e_{i,j}$ obtained from the expression profiles, as well as the average weight $w^m_{i,j}$ obtained from the markers, both computed over the edges $i \to j$ that are part of the gold standard network. The weights $w^e_{i,j}$ and $w^m_{i,j}$ are those obtained on the datasets simulated with the setting configuration 1 (large marker distance, high heritability, and small population size). Despite the high heritability, the tree-based importance values of the genetic markers, as computed in GENIE3-SG-joint, are typically much lower than those obtained from the expression data

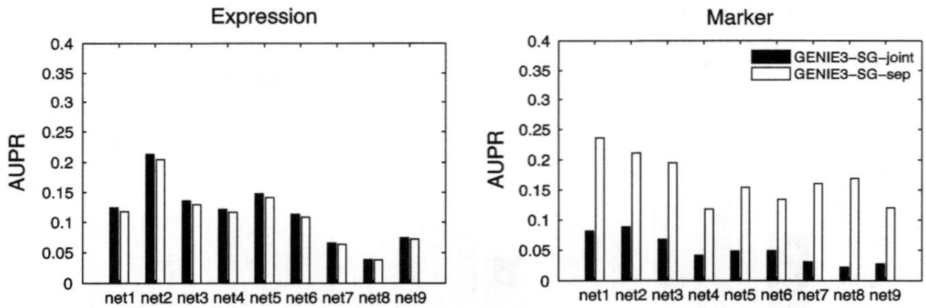

Fig. 5.6 AUPR scores of expressions and markers. This figure shows the AUPR scores obtained when the weight of each edge $i \to j$ is the importance $w^e_{i,j}$ obtained from the expression data (*left*) and the importance $w^m_{i,j}$ obtained from the genetic markers (*right*). The ranking of interactions obtained from the markers is significantly less accurate with GENIE3-SG-joint, compared to GENIE3-SG-sep. The AUPR values of each method were averaged over the eight datasets corresponding to each of the nine networks

Aggregation procedures

For both procedures GENIE3-SG-joint and GENIE3-SG-sep, higher scores are obtained when the importance scores $w^e_{i,j}$ and $w^m_{i,j}$ are aggregated by taking their product rather than their sum (Fig. 5.4), i.e., when we consider that both the genetic marker and the expression of a regulating gene are important for the prediction of the expression of a target gene. This conservative aggregation procedure allows to give a lower weight to a lot of false edges, since many of them can still have a high value of $w^e_{i,j}$ or a high value of $w^m_{i,j}$ without having a regulatory effect (Fig. 5.7). By contrast, high values for both $w^e_{i,j}$ and $w^m_{i,j}$ are obtained only for true edges.

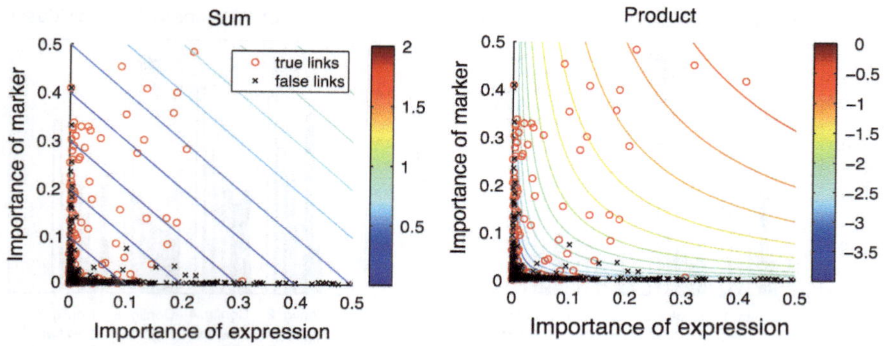

Fig. 5.7 Scatter plot of importances of expressions and markers returned by GENIE3-SG-sep. The red circles correspond to the true edges of the gold standard network, while the black crosses correspond to the false edges (i.e., edges that are not part of the gold standard). In addition, each plot shows the contour lines of the sum and product of $w^e_{i,j}$ and $w^m_{i,j}$ (*left* and *right* figures respectively). The values of the product are shown in a logarithmic scale. These results are those obtained on the first network of 100 genes (Network 100-1), under configuration 1 (high heritability, large marker distance, and small population size). Taking the product of the importance scores allows to give a lower weight to a lot of false edges

5.3.2.4 Influence of Population Size, Heritability, and Marker Distance

Figure 5.8 shows the AUPR scores obtained by the different GENIE3 procedures on the networks of 1,000 genes, for each setting configuration of the simulation runs. As expected, the performance of each method improves when the number of individuals for which data are available increases. The scores also indicate that genetic markers are much more informative than expression data for the inference of the networks when the median heritability as well as the distance between the markers are both high (configurations 1 and 2). In these configurations, only exploiting genetic data ("GENIE3 on markers") results in significantly more accurate predictions than learning from expression data alone ("GENIE3 on expression"). This result is not surprising since a higher heritability means that a higher proportion of the variance of the expression data is actually explained by the genetic markers. Moreover, a higher distance between the markers implies an increased rate of chromosomal crossovers between the different markers, and hence more meaningful multifactorial perturbations (i.e., genetic variations) between the individuals, helping to recover the networks in a more accurate way. By contrast, when the heritability and the marker distance are both small (configurations 7 and 8), expression data are more informative than the markers. In the remaining configurations (configurations 3–6), the performance obtained from gene expression is not very different from the one obtained from genetic markers. Nevertheless, it seems that expression and genetic data contain different and complementary information about the underlying networks, since in all configurations the predictions can be highly improved when both types of data

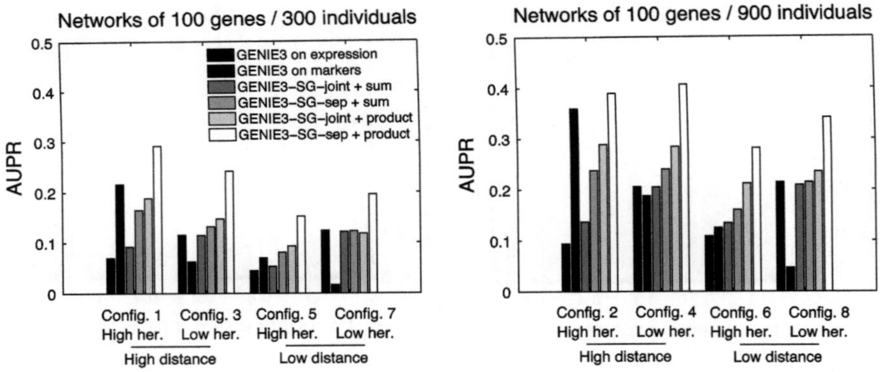

Fig. 5.8 Performances of inference methods for each setting configuration. Predictions can be highly improved when expression and genetic data are integrated, the highest AUPR scores being obtained by GENIE3-SG-sep, using the product of the weights $w^e_{i,j}$ and $w^m_{i,j}$. GENIE3 on expression: weight of edge $i \to j$ is the importance $w^e_{i,j}$ of the expression of gene i as computed in GENIE3-SG-sep. GENIE3 on markers: weight of edge $i \to j$ is the importance $w^m_{i,j}$ of the genetic marker of gene i as computed in GENIE3-SG-sep. Her.: heritability. The AUPR values of each method were averaged, for each configuration, over the three datasets related to the networks of 1,000 genes

are integrated, the best results being achieved by far by GENIE3-SG-sep using the product of the weights $w^e_{i,j}$ and $w^m_{i,j}$.

The PR curves related to the first network of 1,000 genes (Network 1000-1) are plotted in Fig. 5.9, for the configuration 1 (high heritability, large marker distance, and small population size). As an example, the 500 first regulatory links obtained with GENIE3-SG-sep (product) yield a precision of 77 % and a recall of 12 %. Increasing the number of considered edges to 1,000 allows us to recover more true edges (recall equal to 18 %), with however, a decrease in precision (57 %). Limiting the network to the first 200 links allows to keep a precision higher than 90 % (recall equal to 6 %). On the other hand, more than 800,000 links have to be considered to obtain a recall higher than 90 % (precision equal to 0.4 %), which is of course of no practical interest.

5.3.2.5 Direction of the Edges

One interesting feature of the GENIE3 methods is their potential ability to predict directed networks. To assess the ability of each method to predict link directions, we computed the error rate on the direction of the edges, i.e., the proportion of edges $i \to j$ in the gold standard network such that there is no edge $j \to i$ and for which the method wrongly predicts $w_{i,j} < w_{j,i}$. The error rates are shown in Fig. 5.10. Compared to exploiting expression data alone, using information about genetic markers greatly helps for the prediction of the direction of the edges. However, there is no significant difference between the different methods exploiting the markers. As an example, the GENIE3-SG-sep (product) method yields an average error rate of 27 %.

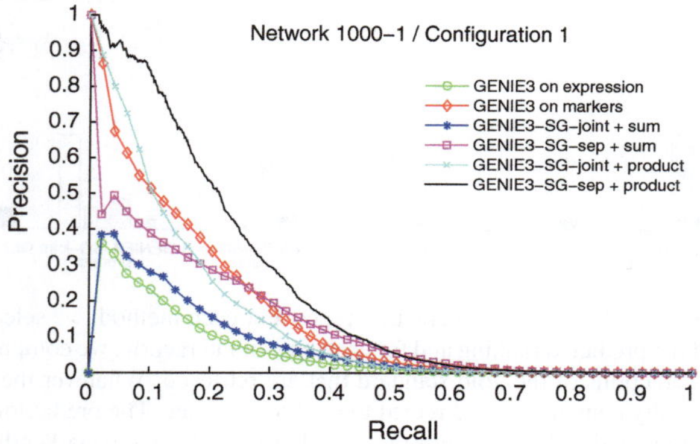

Fig. 5.9 Precision–recall curves. These PR curves were obtained for the first network of 1,000 genes (Network 1000-1), under configuration 1 (high heritability, large marker distance, and small population size). As an example, the 500 first regulatory links obtained with GENIE3-SG-sep (product) yield a precision of 77 % and a recall of 12 %. GENIE3 on expression: weight of edge $i \to j$ is the importance $w^e_{i,j}$ of the expression of gene i as computed in GENIE3-SG-sep. GENIE3 on markers: weight of edge $i \to j$ is the importance $w^m_{i,j}$ of the genetic marker of gene i as computed in GENIE3-SG-sep

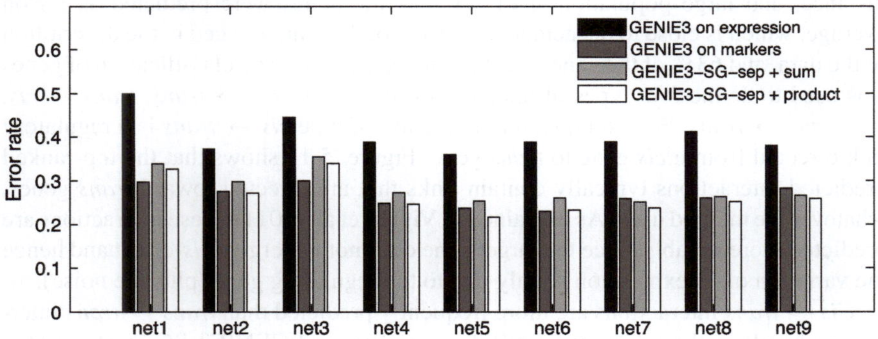

Fig. 5.10 Error rates on edge directionality. Using information about genetic markers greatly helps for the prediction of the direction of the edges. GENIE3 on expression: weight of edge $i \to j$ is the importance $w^e_{i,j}$ of the expression of gene i as computed in GENIE3-SG-sep. GENIE3 on markers: weight of edge $i \to j$ is the importance $w^m_{i,j}$ of the genetic marker of gene i as computed in GENIE3-SG-sep. The error rates of each method were averaged over the eight setting configurations of Table 5.1 corresponding to each of the nine networks

5.3.2.6 Interactions Types

We adopted the same evaluation protocol as in Vignes et al. (2011), and analyzed the performance of the GENIE3-SG-sep (product) method, as a function of the type of interactions. We labeled a gene "*cis*" if the corresponding genetic marker is detected in its promoter region, and "*trans*" if the marker is in its coding region. The gene

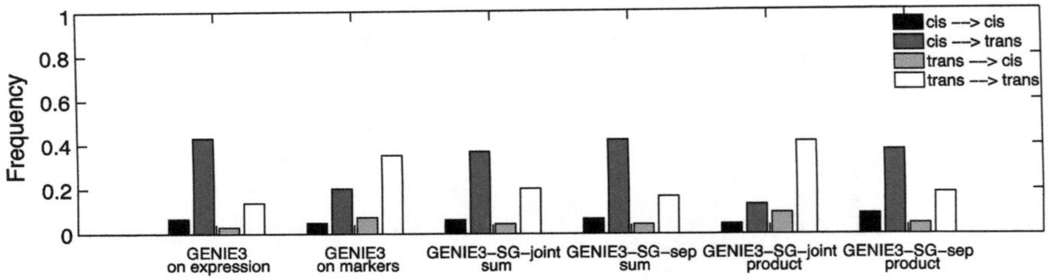

Fig. 5.11 Frequency of the different interaction types. For each method, we selected the first 500 regulatory links of the predicted ranking and for each type of interaction, we computed its frequency among the true interactions of the gold standard that are retrieved. Whatever the method, the top of the ranking typically contains links directed toward *trans* genes. The predictions were obtained from the datasets simulated under configuration 2 (high heritability, large marker distance, and large population size), and the interactions frequencies were averaged over the nine networks

classification was obtained by performing an analysis of variance, as described in Vignes et al. (2011). Genes with a corrected p-value lower than 0.001 were identified as *cis*, and those with an uncorrected p-value higher than 0.1 were identified as *trans*. Using the datasets simulated under configuration 2 (high heritability, large marker distance, and large population size), 23 % of the genes were predicted as *cis* on average, which is close to the actual proportion of 25 % announced in the description of the data, and 63 % of the genes were predicted as *trans*. The classification of genes was used to define four types of interactions: *cis* → *cis*, *cis* → *trans*, *trans* → *cis*, and *trans* → *trans*. For example, an interaction of type *cis* → *trans* is a regulatory link directed from a *cis* gene to *trans* gene. Figure 5.11 shows that the top-ranked predicted interactions typically contain links that are directed toward *trans* genes, whatever the method used. As explained in Vignes et al. (2011), these interactions are predicted more reliably since the target gene does not undergo a *cis*-effect and hence the variation of its expression is only due to the regulating gene (plus the noise).

Cis → *trans* interactions are more frequently predicted than *trans* → *trans* interactions by all methods except GENIE3 on markers and GENIE3-SG-joint with the product. This difference can be explained. In *trans* → *trans* interactions, the impact of the marker of the regulating gene on the expression of the target gene is indeed more direct than in *cis* → *trans* interactions, where the marker only affects the expression of the target gene through the expression of the regulating gene. This leads to higher scores for *trans* → *trans* interactions when GENIE3 is applied on markers only. In GENIE3-SG-joint, the scores of markers and expressions are more balanced for *trans* → *trans* interactions than for *cis* → *trans*, as in *trans* → *trans* interactions both the marker and the expression of the regulating gene are directly and independently affecting the expression of the target gene. This eventually leads to higher scores for *trans* → *trans* interactions when taking the product of marker and expression scores.

5.3.3 *The DREAM5 Systems Genetics Challenge*

5.3.3.1 Description of the Challenge

The Dialogue for Reverse Engineering Assessments and Methods (DREAM) initiative organizes an annual reverse engineering competition that comprises several challenges[5] (Marbach 2012; Prill et al. 2010; Stolovitzky et al. 2009, 2007). We report here our results on the DREAM5 *Systems Genetics* challenge.[6] This challenge concerned the inference of *in silico* regulatory networks from systems genetics data. It was divided into three sub-challenges. The goal of each sub-challenge was to infer five networks from populations of 100, 300, and 999 individuals respectively. Each of the 15 networks contained 1,000 genes and were of increasing connectivity within each sub-challenge. For each individual, expression levels of all the genes were provided, as well as the genotype value of one genetic marker for each gene. All data of the challenge were generated using a preliminary version of SysGenSIM. However, the information about the configuration used to run the simulations was not provided to the challenge participants.

5.3.3.2 Performance of the GENIE3 Methods

Figure 5.12 shows the performances of the different methods. We observe results similar to those obtained on the StatSeq datasets: the performance increases with the number of individuals, better performances are obtained with GENIE3-SG-sep compared to GENIE3-SG-joint, and the product of importance scores $w^e_{i,j}$ and $w^m_{i,j}$ also yields higher AUPR scores than the sum.

5.3.3.3 Comparison with the DREAM5 Best Performer

Figure 5.13 compares, in terms of AUPR scores, GENIE3-SG-sep to the procedure that was used by the official best performing team of the DREAM5 *Systems Genetics* challenge. This procedure is a meta-analysis of different methods, respectively, based on Dantzig regression (Candès and Tao, 2007), LASSO regression (Tibshirani, 1996), and static Bayesian network learning (Friedman et al., 2000). The procedure is described in detail in Vignes et al. (2011). The AUPR scores indicate that our procedure significantly outperforms the meta-analysis for each of the networks.

---

[5] http://www.the-dream-project.org/.

[6] http://wiki.c2b2.columbia.edu/dream/index.php/D5c3.

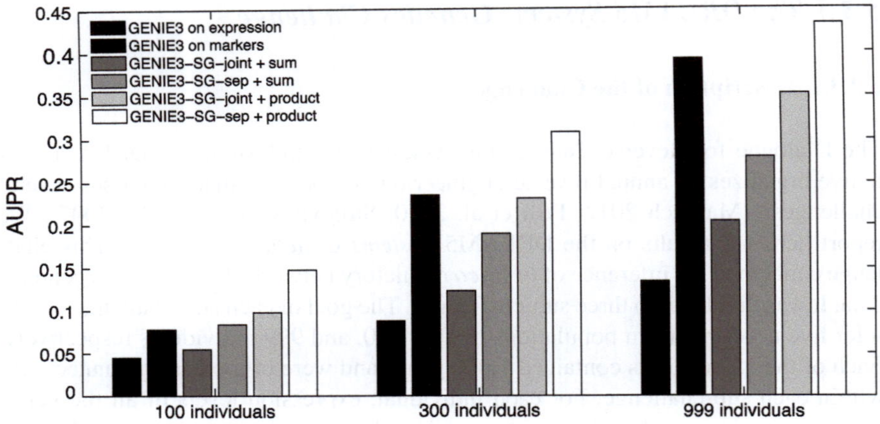

Fig. 5.12 AUPR scores for the DREAM5 Systems Genetics challenge. As for the StatSeq networks, the best predictions are obtained with GENIE3-SG-sep, using the product of the weights $w^e_{i,j}$ and $w^m_{i,j}$. GENIE3 on expression: weight of edge $i \rightarrow j$ is the importance $w^e_{i,j}$ of the expression of gene i as computed in GENIE3-SG-sep. GENIE3 on markers: weight of edge $i \rightarrow j$ is the importance $w^m_{i,j}$ of the genetic marker of gene i as computed in GENIE3-SG-sep. The AUPR values of each method were averaged over the five networks of each sub-challenge

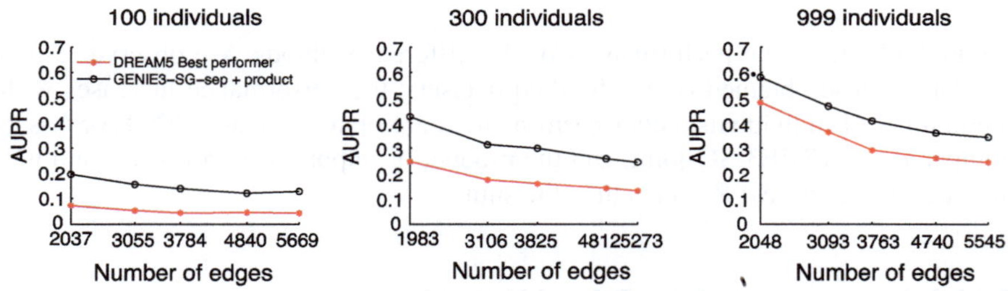

Fig. 5.13 Comparison with the best performer and influence of network density. The GENIE3-gen-sep (product) method outperforms the procedure of the official best performer of the challenge. The performance of both methods, however, decreases when the number of edges in the network increases

5.3.3.4 Influence of Network Density

Besides the study of the effect of the dataset size (number of individuals) on the predictions returned by inference methods, the DREAM5 *Systems Genetics* challenge was also designed to study the effect of the connectivity of a network on the predictions. Figure 5.13 shows the effect of the network density on the predictions. Clearly, in each sub-challenge, the ability of the methods to recover a network tends to decrease as the number of edges in the network increases and regulatory interactions become more complex.

5.4 Discussion

In this chapter, we proposed two procedures, GENIE3-SG-joint and GENIE3-SG-sep, that infer (GRNs) from systems genetics data. Both procedures decompose the problem of inferring a regulatory network of p genes into p different feature selection problems, the goal of each being to retrieve the regulators of one of the genes of the network. Each feature selection problem is then solved by applying a tree-based ensemble method in order to obtain a model predicting the expression of one gene j. In the GENIE3-SG-joint procedure, a single predictive model is learned from expression and genetic data, while in GENIE3-SG-sep, two separate predictive models are learned, one based on the genetic markers and the other based on the expression data. Both methods then compute, for each gene $i \neq j$, two scores $w^e_{i,j}$ and $w^m_{i,j}$, measuring, respectively, the importances of the expression and of the marker of gene i when predicting the expression of gene j. These two scores are then aggregated, by computing either their sum or their product, to obtain a single weight $w_{i,j}$ for the regulatory link directed from gene i to gene j.

The artificial datasets of the StatSeq benchmark were simulated using different setting configurations, i.e., by combining different values of the number of individuals, distance between the markers, and heritability, allowing us to check under which configurations our different methods perform best. Results showed that depending on the marker distance and heritability, genetic markers bring more or less information about the regulatory networks than expression data, and combining the two types of data can be highly helpful for their recovering. GENIE3-SG-sep, using the product of the weight $w^e_{i,j}$ and $w^m_{i,j}$, yields the best performances, whatever the configuration. This method also yields the best performances when recovering the networks of the DREAM5 *Systems Genetics* challenge, and actually outperforms the official best performing algorithm of the challenge.

The StatSeq datasets and the DREAM5 challenge allowed us to make a first evaluation of the performances of our different procedures on systems genetics data. However, these benchmarks are solely based on networks and data that are artificial. As future works, we thus would like to apply our methods on real datasets. Datasets related to various organisms are publicly available, such as the *S. cerevisiae* dataset of Brem and Kruglyak (2005). However, in our different procedures, we assume that each gene whose expression is measured in N individuals is also analyzed for one single genetic marker in each of these N individuals. Unfortunately, this situation is usually not encountered in real datasets. We will thus have to modify our methods in order to deal with missing data, and also to establish a procedure to aggregate the importance scores of different genetic markers related to the same gene.

Finally, although we exploited tree-based ensemble methods, the frameworks of GENIE3-SG-joint and GENIE3-SG-sep are general, and other feature ranking techniques could have been used as well. In the future, we thus plan to apply and compare different ranking techniques, and check which of them permit the best exploitation of expression and genetic data.

Acknowledgments The authors would like to thank Jimmy Vandel for useful discussions on the topic during a visit in February 2011. This work was partially funded by the Interuniversity Attraction Poles Programme (IAP P6/25 BIOMAGNET and IAP P7 DYSCO), initiated by the Belgian State, Science Policy Office, and by the European Network of Excellence PASCAL2. Pierre Geurts is a research associate of FNRS, Belgium. The authors thank the GIGA Bioinformatics platform and the SEGI (University of Liège) for providing computing resources.

References

Aten JE, Fuller TF, Lusis AJ, Horvath S (2008) Using genetic markers to orient the edges in quantitative trait networks: the NEO software. BMC Syst Biol 2:34

Bing N, Hoeschele I (2005) Genetical genomics analysis of a yeast segregant population for transcription network inference. Genetics 170:533–542

Breiman L (1996) Bagging predictors. Mach Learn 24:123–124

Breiman L (2001) Random forests. Mach Learn 45:5–32

Breiman L, Friedman JH, Olsen RA, Stone CJ (1984) Classification and regression trees. Wadsworth International, California

Brem RB, Kruglyak L (2005) The landscape of genetic complexity across 5,700 gene expression traits in yeast. Proc Natl Acad Sci USA 102:1572–1577

Candès E, Tao T (2007) The dantzig selector: Statistical estimation when p is much larger than n. Ann Stat 35:2313–2351

Neto Chaibub E, Ferrara CT, Attie AD, Yandeli BS (2008) Inferring causal phenotype networks from segregating populations. Genetics 179:1089–1100

Chen LS, Emmert-Streib F, Storey JD (2007) Harnessing naturally randomized transcription to infer regulatory relationships among genes. Genome Biol 8:R219

De Smet R, Marchal K (2010) Advantages and limitations of current network inference methods. Nat Rev Microbiol 8:717–729

Friedman N, Linial M, Nachman I, Pe'er D (2000) Using Bayesian networks to analyze expression data. J Comp Biol 7:601–620

Guyon I, Elisseeff A (2003) An introduction to variable and feature selection. JMLR 3:1157–1182

Hastie T, Tibshirani R, Friedman J (2009) The elements of statistical learning: Prediction, inference and data mining. Springer Verlag, Second Edition

Huynh-Thu VA, Irrthum A, Wehenkel L, Geurts P (2010) Inferring regulatory networks from expression data using tree-based methods. PLoS ONE 5:e12776

Jansen RC (2003) Studying complex biological systems using multifactorial perturbation. Nat Rev Genet 4:145–151

Jansen RC, Nap J-P (2001) Genetical genomics: the added value from segregation. Trends Genet 17:388–391

Kulp DC, Jagalur M (2006) Causal inference of regulator-target pairs by gene mapping of expression phenotypes. BMC Genomics 7:125

Li. H, Lu L, Manly KF, Chesler EJ, Bao L, Wang J, Zhou M, Williams RW, Cu i Y (2005) Inferring gene transcriptional modulatory relations: a genetical genomics approach. Hum Mol Gen 14:1119–1125

Li R, Tsaih S-W, Shockley K, Stylianou IM, Wergedal J, Paigen B, Churchill GA (2006) Structural model analysis of multiple quantitative traits. PLoS Genet 2:e114

Liu B, de la Fuente A, Hoeschele I (2008) Gene network inference via structural equation modeling in genetical genomics experiments. Genetics 178:1763–1776

Marbach D, Costello J (2012) C., Küffner, R., Vega, N., Prill, R. J., Camacho, D. M., Allison, K. R., the DREAM5 Consortium, Kellis, M., Collins, J. J., Stolovitzky, G.: Wisdom of crowds for robust gene network inference. Nat Methods 9:796–804

Meinshausen N, Bühlmann P (2006) High-dimensional graphs and variable selection with the Lasso. Ann Stat 34:1436–1462

Meyer PE, Kontos K, Lafitte F, Bontempi G (2007) Information-theoretic inference of large transcriptional regulatory networks. EURASIP J Bioinform Syst Biol 2007:79879

Michaelson JJ, Alberts R, Schughart K, Beyer A (2010) Data-driven assessment of eQTL mapping methods. BMC Genomics 11:502

Pinna A, Soranzo N, Hoeschele I, de la Fuente A (2011) Simulating systems genetics data with SysGenSIM. Bioinformatics 27:2459–2462

Prill RJ, Marbach D, Saez-Rodriguez J, Sorger PK, Alexopoulos LG, Xue X, Clarke ND, Altan-Bonnet G, Stolovitzky G (2010) Towards a rigorous assessment of systems Biology models: the DREAM3 challenges. PLoS ONE 5:e9202

Saeys Y, Inza I, Larranaga P (2007) A review of feature selection techniques in bioinformatics. Bioinformatics 23:2507–2517

Schadt EE, Lamb J, Yang X, Zhu J, Edwards S, Guhathakurta D, Sieberts SK, Monks S, Reitman M, Zhang C, Lum PY, Leonardson A, Thieringer R, Metzger JM, Yang L, Castle J, Zhu H, Kash SF, Drake TA, Sachs A, Lusis AJ (2005) An integrative genomics approach to infer causal associations between gene expression and disease. Nat Genet 37:710–717

Stolovitzky G, Monroe D, Califano A (2007) Dialogue on Reverse-Engineering assessment and methods: the DREAM of high-throughput pathway inference. Ann NY Acad Sci 1115:11–22

Stolovitzky G, Prill RJ, Califano A (2009) Lessons from the DREAM2 challenges. Ann NY Acad Sci 1158:159–195

Strobl C, Boulesteix A-L, Zeileis A, Horthorn T (2007) Bias in random forest variable importance measures: illustrations, sources and a solution. BMC Bioinform 8:25

Tibshirani R (1996) Regression shrinkage and selection via the Lasso. J R Stat Soc Ser B 58:267–288

Vignes M, Vandel J, Allouche D, Ramadan-Alban N, Cierco-Ayrolles C, Schiex T, Mangin B, de Givry S (2011) Gene regulatory network reconstruction using bayesian networks, the Dantzig selector, the Lasso and their meta-analysis. PLoS ONE 6:e29165

Zhu J, Wiener MC, Zhang C, Fridman A, Minch E, Lum PY, Sachs JR, Schadt EE (2007) Increasing the power to detect causal associations by combining genotypic and expression data in segregating populations. PLoS Comput Biol 3:e69

Chapter 6
Extending Partially Known Networks

Pegah Tavakkolkhah and Robert Küffner

Abstract Besides experimental techniques, computational inference approaches have contributed to the reconstruction of gene regulatory networks in model organisms. Particularly successful are supervised approaches that take the known regulatory interactions and gene expression data into account. However, they have not yet been applied to individuals genotyped by systems genetics data, where genetic polymorphisms are the major source of variation in gene expression profiles. We apply a supervised inference framework to expression datasets, genotype information, and the known gene regulatory interactions that are generated in a standardized setup by the SysGenSim software. We confirmed in this setup that supervised approaches exploiting the known interactions perform better than pure expression-based methods as well as methods exploiting expression data and genotype information. The performance of supervised methods was robust with respect to parameterization and data pre-processing. Furthermore, whether or not the genotype information was explicitly used influenced the performance of supervised approaches only little. We also analyzed differences between real and artificial data and setups to assess the chances of a successful inference in real systems. Due to reasons discussed in this chapter, several extensions of supervised approaches that considerably improve performance on real data were not effective in the SysGenSim case. Our thorough comparison between real and artificial setups suggested that the application of supervised approaches to real systems might be more robust and straightforward in comparison to current unsupervised approaches. In particular, as real genotypes likely are more complex and cause more versatile responses, the finding that supervised approaches are not dependent on the explicit representation of genotype information might prove of advantage.

P. Tavakkolkhah · R. Küffner (✉)
Department of Informatics, Ludwig-Maximilians Universität, Amalienstr. 17,
80333 München, Germany
e-mail: robert.kueffner@bio.ifi.lmu.de

P. Tavakkolkhah
e-mail: pegah@bio.ifi.lmu.de

A. de la Fuente (ed.), *Gene Network Inference*,
DOI: 10.1007/978-3-642-45161-4_6, © Springer-Verlag Berlin Heidelberg 2013

6.1 Introduction

Inference of gene regulatory networks from high-throughput data provides new insights into the regulatory mechanisms that control the expression of genes in cells. Gene regulatory networks control the response of the cell to local and environmental input signals by regulating the expression of genes. Considering that genes are not necessarily regulated by single transcription factors (TFs), but different combinations of regulators might be involved, the theoretical possible number of combinatorial regulations might impede the full experimental determination of all interactions. In addition, TFs are activated under specific experimental conditions that play an important role in detecting potential interactions. As a result, employing computational approaches can complement our knowledge of regulatory networks and a range of different kinds of data and methods have been devised and employed to address the regulatory network inference problem.

The inference process is based on analysis of the system's behavior to perturbations (De Smet and Marchal 2010). Environmental perturbations, for example, manipulate experimental conditions such as heat shock, or chemical stress. The transcriptome data has been the main source for analyzing such perturbations or predicting gene regulatory interactions, mainly because measurement of transcription (mRNA level) is more feasible compared to proteomic and metabolic datasets. It has to be taken into account, though, that alternative mRNA splicing (increasing the number of possible proteins compared to the number of protein-encoding genes) and post-translational mechanisms may increase the diversity of the mRNA response, and that many of these mechanisms are not immediately observable on the level of mRNA profiling. For instance, proteins enter complexes with other proteins or various other molecules including RNAs molecules, which are required for their specific functions. Thus, describing the behavior of the network only from the perspective of the mRNA expression level ignores all other mechanisms involved in molecular cell regulation.

Other types of experimental datasets informative for the reconstruction of networks can be obtained on an organism-wide scale that include Chromatin Immuno-Precipitation (ChIP, e.g., as ChIP-on-chip, i.e., ChIP combined with microarray technology or ChIP-seq, i.e., ChIP combined with sequencing; Zheng et al. 2010; Abdulrehman et al. 2011), or protein–protein interactions. For example, ChIP identifies the binding site of a particular TF (TF binding sites, TFBS) on the entire genome, and particularly at the promoter region of putative target genes (TG). TFBS reveals potential TF:TG regulatory effects, since TF binding might lead to transcription of the gene.

Genetic perturbations, on the other hand, involve techniques such as gene deletion (knock-out, KO; Hu et al. 2007) or overexpression (OE; Chua et al. 2006), frequently combined with mRNA profiling experiments, to detect regulatory dependencies. Such perturbations are of particular interest as given regulators can be directly targeted to identify their downstream effects and perturbations therefore can reveal regulatory dependencies between perturbed gene and its targets. The impact of perturbations can be profiled via microarrays or next generation sequencing techniques to identify repression or induction of expression levels (according to the type of the interaction) of genes regulated by the corresponding TF. However, it should be noted that the change in behavior is not only limited to direct targets of the TF in question, but also pathways indirectly regulated by the perturbed TF might also show distinct signals. Although informative, the cost of (single- or multi-factorial) KO and OE experiments can become prohibitive.

Genetic polymorphisms are particular examples of multifactorial in vivo perturbation (Jansen et al. 2001). A polymorphism is due to the crossing of genetically diverse parent strains and further crossings of their children for several generations, thus effectively randomizing the distribution of alleles. The individuals of the final population, which is known as recombinant inbred line (RIL) segregating population, comprise a wide diversity of genetic information in respect to parent sequences. The variations between RILs represent random multifactorial perturbations that can further affect the cell phenotypes. The study of these variations between RILs and their direct/indirect effect on cell traits is known as "genetical genomics" or "systems genetics." Analysis of genetic markers (in the genotype data used here each gene is characterized by just two states) and perturbed phenotypes (here, expression data) for each individual are two common steps to identify the potential causal dependencies between genetic variations and cell behavior (Loh et al. 2011; Vignes et al. 2011). While such polymorphisms are more cost and time effective, from an experimental point of view, than targeted genetic perturbations as they are already present in population of interest or easy to come by, they are more difficult to interpret as they exhibit a multifactorial response that is difficult to connect to a single cause.

Nevertheless, as genetic polymorphism is inherent in most expression profiles, it has, explicitly or implicitly, been exploited for network inference. The conceptually simplest types of approaches (although they can be demanding algorithmically and/or computationally) are approaches that rely on expression data alone to infer networks (De Smet and Marchal 2010). There are several strategies to design such expression-based inference approaches. Module inference methods such as Stochastic LeMoNe (learning module networks) rely on clustering genes according to their response to specific conditions, i.e., genes with similar behaviors tend to be clustered in the same sets. This approach defines regulatory programs that explain the behavior of a set of genes and as a result the possible number of interactions is drastically reduced. On the other hand, direct inference methods such as CLR (context likelihood of relatedness; Faith et al. 2007) and ARACNE and ANOVA-based regressions (Margolin et al. 2006; Küffner et al. 2012) consider each gene individually.

Another inference strategy is to use part of the information that we want to predict as input knowledge, which is known as supervised inference in machine learning terminology (Bleakey et al. 2007; Mordelet and Vert 2008). Supervised methods use part of the gold standard, in addition to other data sources, to train a model/classifier, which can be further used to classify genes into interacting and non-interacting classes. Using gold standard as training data is feasible, since many TF:TG interactions have been curated in model organisms so far (e.g. RegulonDB for E. coli or YeastRACT for yeast). Furthermore, it has been shown that cross-species knowledge transfer can suggest interactions known in one species to be present in another based on sequence homology (Reference!!).

Inference methods can be further categorized as integrative and non-integrative (De Smet and Marchal 2010). While non-integrative methods are only based on using expression data (and perhaps part of the known gene regulatory interactions as in case of supervised approaches), integrative approaches use other data sources as well. Integrative methods aim to combine other datasets with expression data to predict new interactions. Using other data sources is expected to not only increase the validity of the inferred network, but also to provide a more comprehensive picture of the network, including not only genes but also proteins and metabolites. The type of data most commonly used in this context are TFBS (Ernst et al. 2008) represented and employed for predicting putative TF:TG interactions in the form of positional weight matrices (PWMs). A second source of information that has been utilized for inference are chromatin profiles (Ernst et al. 2011).

Network inference is not limited to reconstruction of gene regulatory networks, but other domains such as protein networks (Bleakley et al. 2007), metabolic networks (Vert and Kanehisa 2003), or protein functions (Lanckriet et al. 2004; Mostafavi et al. 2008) are also of interest as these different types of networks are tightly interlinked and mutually influence each other. Protein networks are used to describe interacting proteins, while protein functions represent associations between proteins and biological concepts such as biological processes. Protein networks and protein functions are often employed for mutual inference as for instance the networks could be used to predict functions of unknown proteins. Reconstruction of protein networks is challenging, not only due to the high complexity of the problem but also due to the fact that these networks are largely unknown. Another issue is due to the vast number of post-transcriptional modifications of proteins that might imply functional switches, which in turn might correspondingly change the role of affected proteins in the network. Metabolic networks are an example of protein networks, where enzymes represent metabolic reactions that transform different chemical compounds called metabolites. Metabolic reactions can thus be represented as bipartite graph (with enzymes and metabolites as nodes) or as common graph where an edge is implied for instance between two enzymes catalyzing two successive reactions. Computational approaches for the inference and simulation of networks frequently address only a restricted sub-domain of molecular networks and their regulation (Hecker et al. 2009).

In this work we explore the application of supervised approaches to reconstruct gene regulatory networks from known gene regulatory interactions and information on genotypes that were used by the simulator SysGenSim (SGS; Pinna et al. 2011) to generate artificial expression datasets. As central component we apply and extend the approach of SIRENE (Mordelet and Vert 2008), that has, to our knowledge, not been applied to integrate genotypes in the form of systems genetic data. Based on single and combined datasets in the provided SGS compendia, we derive several processed datasets containing information on expression, interactions, genotypes, and their combinations. We thus explore and compare the performance between the original SIRENE approach as well as several supervised and unsupervised variants thereof to determine the utility of genotype information. We further devise a simple unsupervised inference approach based on Wilcoxon's test that exploits genotype information but neglects known gene regulatory interactions. Furthermore, we evaluate the effects of SGS and SIRENE parameters on prediction performance. We consider this work as a pilot study to assess the feasibility of SIRENE and extensions to enable its application on real expression and systems genetics data. Toward this goal, we also evaluate differences between real and simulated datasets to assess the chances for a successful application to real data.

6.2 Methods

6.2.1 Basic Inference Setting and Notations

The majority of gene regulatory network inference approaches analyze expression data to detect regulatory relationships between regulators $r \in R$ such as transcription factors (TFs) and regulated target genes (TGs) $g \in G$. Here the set of regulators R is equivalent to the set of genes G. Usually, all possible pairs (r, g) are examined and scored to rank putative interactions. Three distinct types of data are generated by the software SysGenSim to facilitate the present work: (i) a gene expression data matrix D, (ii) a matrix of genotypes T responsible for the observed changes in gene expression, and (iii) a label matrix consisting of the known gene regulatory interactions that are derived from a predefined gold standard S.

Inference of interactions is based on expression data that is represented by a matrix of $|G|$ genes $\times$ $|E|$ expression measurements. In the present case, different gene expression measurements are generated by the software SysGenSim by a simulation of the effects of genotype variants (gene mutations such as single nucleotide polymorphisms, SNPs). Mutations can be located in the promoter ('cis': influencing the expression of the associated gene and its targets) or in the coding sequence ('trans': influencing target expression only). SysGenSim generates data for $|E|$ recombinant inbred lines (RILs) by recombination. As SysGenSim considers a single locus for mutations per gene that can assume two states '0' and '1', an individual RIL can be specified by a vector of size $|G|$ representing the two mutation states as Boolean values for each gene.

Subsequently, inference results are evaluated by comparing the method-specific confidence scores assigned to putative interactions against a gold standard of known interactions. A gold standard is represented as a matrix S of $|R|$ label vectors of size $|G|$. For each of the regulators $r \in R$, label vectors assign '1' if the transcription factor r regulates a corresponding gene $g \in G$ and '0' otherwise. Method evaluation results in a high performance if the method assigns higher scores to gold standard interactions than to TF:TG pairs that are not regarded as interactions by the gold standard.

Basic performance evaluation
To evaluate the performance of inference, we ranked all predictions (r, g) by decreasing score and computed the area under the ROC curve (AUC) and the area under the precision recall curve (AUPR). The AUC is equal to the probability that a predictor will rank a randomly chosen positive instance higher than a randomly chosen negative one. AUPR measures the average of precision (the number of true positives with a score higher than a threshold) over all thresholds. AUC and AUPR are computed from sensitivity, specificity, and precision values. Sensitivity, also known as recall, measures the probability that the label of a positive sample is correctly identified. Precision and specificity address similar questions but are calculated differently. Precision measures the percentage of true positives that are correctly predicted. Specificity, on the other hand, represents the probability that a negative sample is correctly predicted as negative.

6.2.2 Supervised Inference of Gene Regulatory Interactions

Most approaches for the inference of gene regulatory interactions exclusively rely on the information provided by the gene expression matrix and use the gold standard matrix for evaluation only. In contrast, supervised approaches such as SIRENE (Mordelet and Vert 2008) exploit a subset of the known gene regulatory interactions (GRIs) as labels or prediction targets during the training phase. SIRENE is based on supervised machine learning approaches to train so-called local models, i.e., separate models are trained for all regulators. A model is trained for a given regulator $r \in R$ by selecting the corresponding label vector from the gold standard S. Thus, a given local model can only be used to predict novel targets for the corresponding regulator. If no targets are known for a given regulator, local models cannot be trained and new targets cannot be predicted. Although this is a drawback of local models, its impact will likely be reduced in the future as more and more TF binding experiments are currently performed (ENCODE reference!!). Local models thereby represent the problem of network inference as $|R|$ binary classification tasks, each predicting targets for a single regulator r. Measurements contained in the expression data matrix D are considered as features, i.e., training and prediction is based on $|E|$ features to describe each gene g. Such a binary classification task can be addressed by many machine learning methods. Following the SIRENE approach, we apply support vector machines (SVMs). This setting is denoted as SVC(D) in Fig. 6.1a. In order to evaluate such supervised inference approaches, a 2-times cross-validation (2-fold CV) setup is employed. Half of the gold standard interactions are used for training while the other half is used for evaluation. Subsequently, training and evaluation sets are swapped so that a complete set of $|G| \times |\mathrm{R}|$ scores are obtained. Training and evaluation sets are generated for each regulator in a stratified way, i.e., each set contains the same number of true targets of that regulator according to the gold standard interactions. This 2CV procedure is repeated thrice to obtain more robust performance estimates.

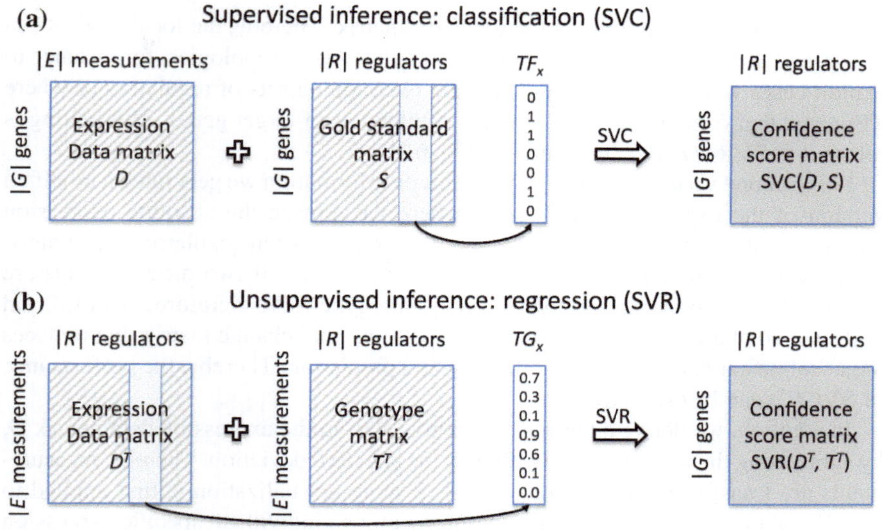

Fig. 6.1 Comparison of inference settings. We implemented various inference settings based on two different formulations. The supervised variant **(a)**, support vector classification (SVC), utilizes the gold standard *S* of gene regulatory interactions in contrast to the unsupervised variant **(b)**, support vector regression (SVR). Feature matrices (hatched) are required for training and prediction, while label vectors (*white*) contain prediction targets that are extracted and thus deleted from the respective feature matrices. Both formulations result in confidence scores (*grey*). Additional variants stem from utilizing different subsets of the available data. For instance, features are defined from expression data only (SVC (*D*)) or from both expression data and the gold standard (SVC(*D*, *S*)). Such a combination ('+' sign) involves the concatenation of several matrices with mutually matching rows, e.g., both of the blue matrices shown in (**a**) have $|G|$ rows that correspond to genes in the same order. Note that in the present application, the set of regulators *R* is equivalent to the set of genes *G*. The set of measurements *E* corresponds to the number of recombinant inbred lines (RILs)

6.2.3 Variants of the Basic Approach

The conventional SIRENE approach relies on a matrix representing the gene expression data and label vectors representing the gold standard network of regulatory interactions. As first level of variants, we evaluate the influence of different SVM parameter settings and kernels. If not mentioned otherwise, we use default parameters with linear kernels. In this study, we also considered a number of additional variants to derive additional features (i) from the genotype information or (ii) from the gold standard.

A first variant derives additional features from the gold standard matrix *S* and thus explicitly represents the known gene regulatory interactions. It is implemented as an extension of the feature matrix that concatenates the expression data matrix *D* and the gold standard matrix *S*. This extension is particularly straightforward as the structures of the two matrices match, i.e., $|G| \times |E|$ in case of the expression data matrix and $|G| \times |R|$ in case of the topology matrix. Thereby, the local models can access and utilize information from non-random network topologies, for instance to exploit cases where genes are regulated by overlapping sets of regulators or where TFs (e.g., that form a complex) regulate similar sets of target genes. This setting is referred to SVR (*D, S*) as indicated by Fig. 6.1b.

For a second variant exploiting the genotype information we generated a modified version of the expression data matrix where we replace the absolute expression measurements D by fold changes C. As we examine each regulator r separately, we expect that fold changes are particularly informative if two measurements are compared to regulator r in genotype 0 versus genotype 1. We therefore select 150 and 300 such comparisons to generate a corresponding fold change matrix that replaces the absolute expression matrix in training and prediction. Thereby, the performance of SVC(*C*) can be estimated.

In addition, we derive a normalized matrix N from the expression data matrix D, by subtracting the average and dividing by the standard deviation. Thereby, measurements are transformed into z-scores. This z-score normalization is first applied to the $|E|$ columns (to normalize the measurement, i.e., individual, specific expression range of measurements) and then to the $|G|$ rows (to normalize the gene-specific expression range) of the matrix D.

The genotype matrix provides information on the generation of the expression data matrix and has the same $|G| \times |E|$ layout. Given that genotypes are generated by random permutation, there should be no co-dependencies between the genotype and network topology and a representation SVC(*D*, *T*) might thus not be immediately informative for the inference of the network topology. We therefore created an intermediate representation displayed in Fig. 6.1b by transposing the genotype (i.e. T^T) and/or expression data matrices (i.e. D^T). This representation is used to preprocess and combine expression data and genotype before training. Instead of classification, we employ support vector regression with linear kernels to predict expression levels of target genes based on either the expression levels of the TFs, the genotype information, or both. Please note that this formulation of the problem resembles the formulation chosen by the Genie3 approach (Huyunh–Thu et al. 2010). Linear SVMs or SVRs trained by optimizing SVM coefficients (α weights) can be converted into TF specific weights by an α-weighted linear combination of the support vectors. This yields SVR (T^T), i.e., a further matrix of a $|G| \times |R|$ layout that can be used for training, e.g., via SVC(*D*, SVR(T^T)).

6.2.4 Unsupervised Inference

For the purpose of method comparison we implemented some unsupervised inference approaches that do not exploit the gold standard interactions. Similar to supervised inference, unsupervised approaches result in matrices of a $|G| \times |R|$ layout containing weights that represent the confidence that a certain gene g is regulated by the corresponding regulator r. Generally, these confidence score matrices can serve two purposes. In the first case, these result matrices are evaluated against the gold standard *S* in the same fashion as described above in the context of supervised inference methods. This constitutes unsupervised inference approaches. In addition, these result matrices can be employed to extend feature matrices in supervised approaches. This illustrates that the support vector regression approaches such as SVR(T^T) or SVR(D^T) can extend feature matrices as described in the previous section and at the same time constitute stand-alone unsupervised approaches.

An additional unsupervised approach is based on the assumption that the targets of a given regulator *r* show significantly different responses if *r* has genotype 0 versus genotype 1. For each regulator *r*, we generate two distributions that take into account all measurements where *r* assumes genotype 0 and 1 respectively. The statistical significance of the difference between these two distributions can be calculated via Wilcoxon's rank sum test. As this approach utilizes both expression data and genotypes we refer to it as WC(*D*, *T*), which again yields a matrix of $|G| \times |R|$ confidence scores.

6.3 Results

6.3.1 Performance of Inference Settings

The standard SIRENE is the application of support vector machines to the expression data matrix as illustrated as SVC (*D*) in Fig. 6.1a. This is the starting point for our analysis and we will explore this setting and several variants (that will be denoted accordingly) on provided SGS data. Figure 6.2a depicts the results of this standard analysis for each of the 72 provided networks. The performance is robust between networks of the same size (100 genes, 1,000 genes, 5,000 genes), while inference on larger networks exhibits a slightly improved performance. In the following, we will by default refer to the networks of 1,000 genes (and averaged measures) if not stated otherwise.

In Figs. 6.2b and 6.3, we show the influence of the choice of input data on method performance. As explained in the method section, SVC generally refers to supervised inference while SVR and WT correspond to unsupervised settings that neglect the information on known interactions contained in the gold standard *S*. The argument *X* of the methods SVC(*X*) or SVR(*X*) denotes the (set of) input datasets, each of which is represented in the form of a matrix. The output of methods is also a matrix as indicated in Fig. 6.1. Predominant input data matrices store expression data (*D*), genotypes (*T*) and the gold standard of known interactions (*S*). From the expression data matrix *D*, we derive two additional matrices, a fold change matrix *C* and a normalized matrix *N*. The matrix *C* has the same size as the expression matrix ($|G| \times |E|$). *E* fold changes for a given gene are calculated from *E* pairs of expression values that are randomly chosen, so that they have opposite mutation types.

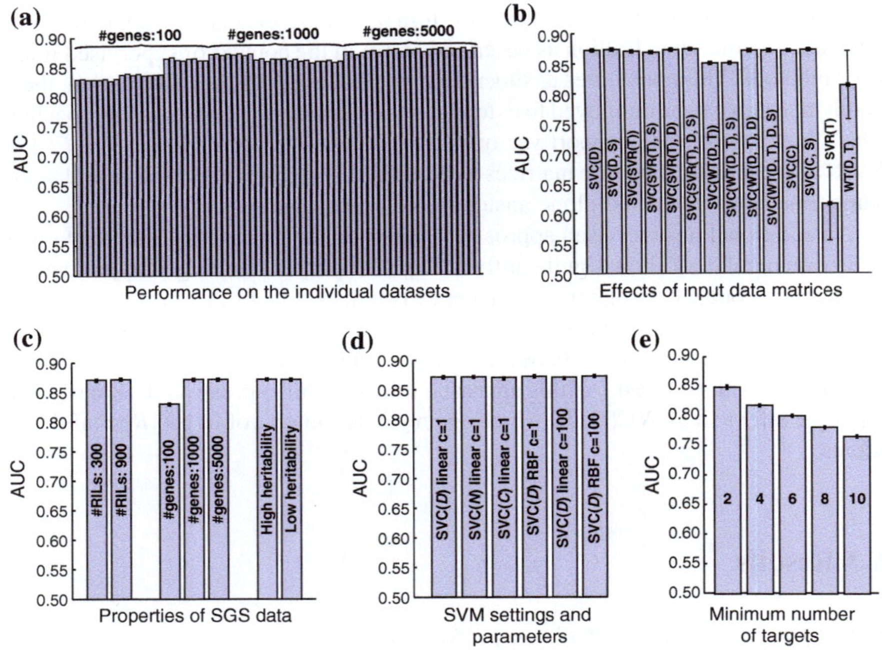

Fig. 6.2 Performance of inference settings (AUC). **a** shows the performance on each of the 72 provided datasets. In (**b**), the effect of different input data matrices on the performance of supervised and unsupervised approaches is depicted. See main text for more information on the method and dataset nomenclature. (**c**) Analyzes how different SysGenSim (SGS) settings influence the generation of expression data. (**d**) Depicts different SVM settings and parameters. In (**e**), the inference is restricted to TFs with a minimum number of target genes. If not noted otherwise, average results are depicted for the standard approach, SVC(D), based on datasets with 1,000 genes

Thus, Figs. 6.2b and 6.3 depict the results of 14 method variants incorporating different combinations of these input datasets.

The best performance of 87–88 % AUC is obtained if matrix D or one of the derived matrices C or N is used as an immediate input to SVC. Additional input datasets such as the gold standard S or the processed genotype T do not substantially increase performance. On the other hand, we observed a slight (non-significant) reduction in performance if only matrices are deployed that have been processed by one of the unsupervised steps (SVR or WT). Note that while the preprocessing has been performed via unsupervised steps, the processed data has been incorporated into a supervised setting, hence SVC(SVR(T)) or SVC(WT(D, T)). The lowest performance is obtained if unsupervised settings are immediately evaluated as SVR(T) or WT(D, T). Here, Wilcoxon's test performed substantially better than a corresponding SVR setup. These general trends are confirmed by the AUPR evaluation in Fig. 6.3 with the exception that the gold standard S led to an increase in the AUPR value.

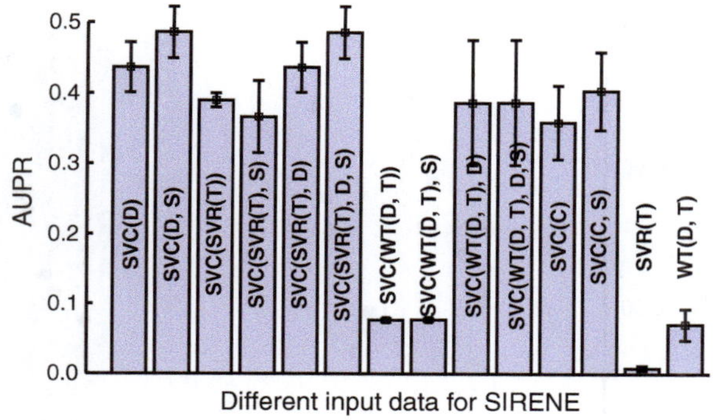

Fig. 6.3 Performance of inference settings (AUPR). The same tests as depicted in Fig. 6.1b are shown here with respect to the area under the precision–recall curve (AUPR)

The performance of SVR(*D*) is compared to the performance of WT(*D, T*) on the 1,000 gene networks in Fig. 6.4. This demonstrates that the performance of the supervised SVR(*D*) in the range of 85–88 % AUC is much more robust than the unsupervised WT(*D, T*) performance exhibiting a much broader range of 71–88 %. Furthermore, this figure depicts 6 datasets of a particularly low WT(*D, T*) performance. According to the SGS annotation, these 6 datasets are characterized by a high biological variance and a low heritability (a term coined by the challenge organizers to describe the chance that markers are inherited, Pinna et al. 2011), which apparently impedes unsupervised inference.

As shown in Fig. 6.2c, different SGS parameters have little influence on performance, confirming the findings in Fig. 6.2a that the performance on 100 gene networks is slightly reduced. Similarly, SVM settings and parameters had almost no impact (Fig. 6.2d).

Generally, supervised local models can only be applied for TFs with known targets. Therefore, we restricted the network to TFs with a preselected minimum number of targets. Requiring a higher minimum consequently decreases the number of predictions. That the performance decreases if more targets are required per TF is a side effect of evaluation, as then more interactions count as 'not predicted,' which reduces the value of AUC. This trend is reversed if only predicted interactions are scored (not shown).

6.3.2 Properties of SGS Data

Here, we explore several SGS properties and parameters, including (i) the effect of TF genotype changes on the TF targets and (ii) the effects of chromosomal proximity.

Figure 6.5 shows that only a subset of target genes (here TG15) is visibly affected by genotype changes in their regulating TFs. Interactions involving targets such as TG15 are easy to detect by unsupervised approaches. For instance, the depicted

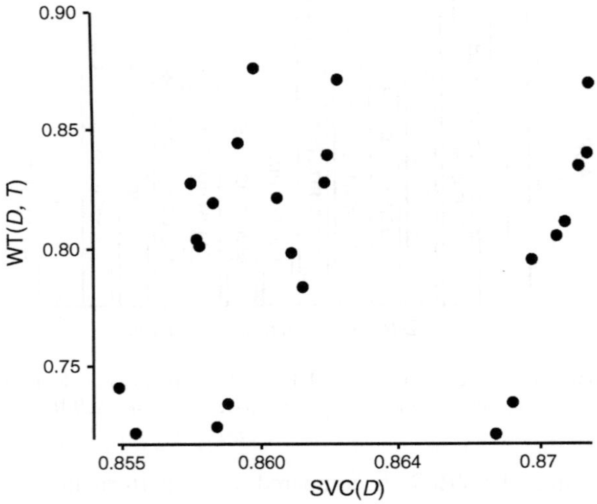

Fig. 6.4 Supervised versus unsupervised performance. This scatter plot compares the AUC performance of two approaches on the 24 datasets with 1,000 genes: (i) a supervised approach exploiting known interactions that is based on support vector classification SVC(*D*), and (ii) an unsupervised approach exploiting genotype information that is based on Wilcoxon's test WT(*D*, *T*). The supervised approach performed better in the examples shown below the diagonal (*dashed line*)

distribution shift is precisely the property that is quantified by the unsupervised Wilcoxon's test WT. In contrast, other target genes such as TG42 show little or no response at genotype changes and are thus more difficult.

We further examined the influence of chromosomal proximity on inferability. If two TFs x and y are located close to each other on the genome, they are subject to linked genetic polymorphisms. In such cases, genotype changes are not informative for the distinction whether a given target gene is regulated by x or y. We examine this by testing the association between genotype changes in a given TF and shifts in the distribution of expression of a given target gene by WT as described in the previous paragraph. In a subsequent step, we extend this to measure the association between a TF and a set of target genes G. Therefore, we employ the AUC to compare the WT scores for genes in G to the WT scores of the remaining genes. Here, an AUC of 1 means that all genes in G experience stronger shifts from genotype changes in a TF than all remaining genes not in G. This is analyzed in Fig. 6.6 where genotype changes in all possible TFs x are analyzed regarding their association with expression changes in targets of TF y (TF13 in Fig. 6.6).

Indeed, genotype changes in TF13 have a strong effect on its targets as demonstrated by the high AUC value of 97 %. However, genotype changes in TFs located close to TF13 on the genome also exhibit a similarly strong association although they neither regulate TF13 nor its targets. Here, differences in genetic polymorphism are correlated with chromosomal distance: nearby TFs experience similar genotype distributions and are thus hard to distinguish based on systems genetics data.

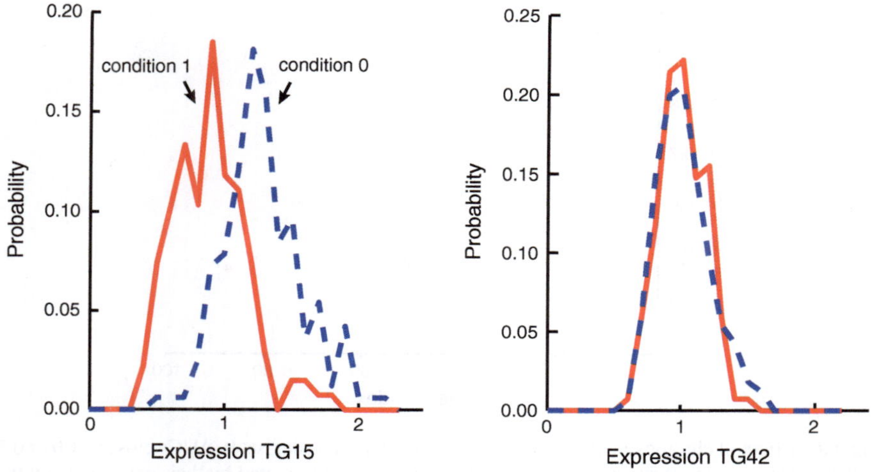

Fig. 6.5 Effect of regulator genotypes on target genes. This plot depicts the distributions of expression levels for two target genes of regulator TF1, namely TG15 and TG42. The switch of genotype 0 (*dashed*) to genotype 1 (*solid*) in regulator TF1 has a noticeable impact on TG15, but not on TG42. Such effects can be exploited for unsupervised inference, e.g., by Wilcoxon's test (see main text)

We indicated two other TFs (TF24 and TF52) in Fig. 6.6 that are the only regulators of TF13, but nevertheless, seemingly exhibit a much lower influence on TF13's targets.

6.3.3 Comparison Between Real and SGS Datasets

In Fig. 6.7 we explore the link between co-regulation and correlation. Each line compares the correlation distribution histograms in units of probability between arbitrary and co-regulate gene pairs. Thus, positive or negative probability differences indicate ranges of correlation with an increased or decreased respectively, likelihood of observing gene pairs regulated by the same TF. The strong negative difference at a correlation of 0 indicates that co-regulated gene pairs are unlikely to exhibit no correlation at all. On the other hand, the positive difference at positive correlation denotes an enrichment of co-regulated genes. Qualitatively, this behavior is shown by SGS and real data while the strength of this effect is substantially more pronounced in case of the SGS data.

We furthermore analyzed pairs of TFs with respect to overlapping sets of target genes in Fig. 6.8. The real networks clearly show much stronger overlaps indicating TF complexes where several TFs regulate (almost) the same targets. We compare this effect to the correlation between pairs of TFs. Interestingly, TFs with strongly overlapping sets of targets exhibit an increased correlation in their expression profiles in case of *E. coli* data. This could indicate that functionally related TFs tend to be

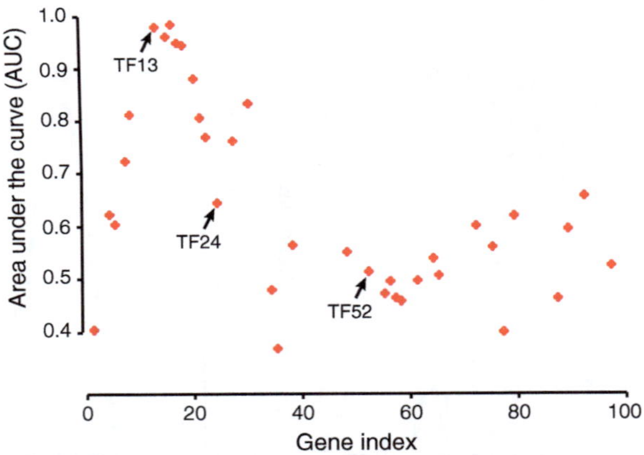

Fig. 6.6 Effect of chromosomal proximity. In this plot based on dataset 9, AUCs close to 1.0 or 0.5 indicate strong or no effects, respectively, of the depicted TFs (denoted by their gene index on the abscissa) on the target genes of TF13. TFs that are located close to TF13 on the genome are subject to similar genetic polymorphisms and their transcriptional effects are thus hard to distinguish. TF24 and TF52 are highlighted that themselves regulate TF13, but have only little effects on target genes of TF13

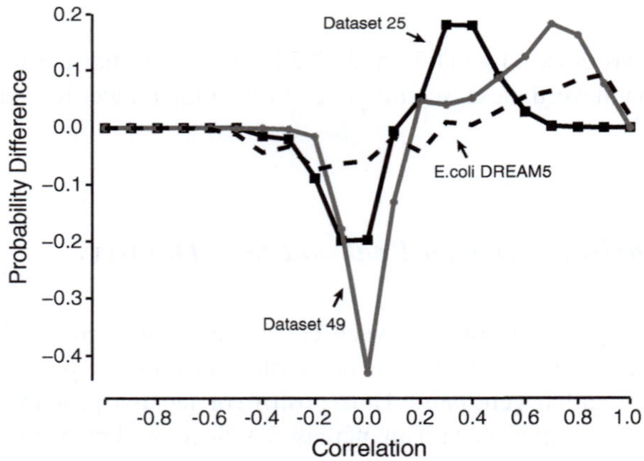

Fig. 6.7 Correlation versus causal relationships. Plotted here is the difference (i)–(ii) between two histograms of normalized probabilities. The two underlying histograms show (i) the correlation between gene pairs regulated by the same TFs and (ii) the correlation between arbitrary gene pairs. Depicted are two SGS derived datasets (*solid*) and one *E. coli* dataset (dashed)

co-regulated as well. Such an effect is not apparent in the SGS data and networks. This suggests that correlation of TFs might be a feature that could be informative in real but not artificial inference settings. Also, the lack of strong target set overlaps offers a potential explanation on why the utilization of the network topology

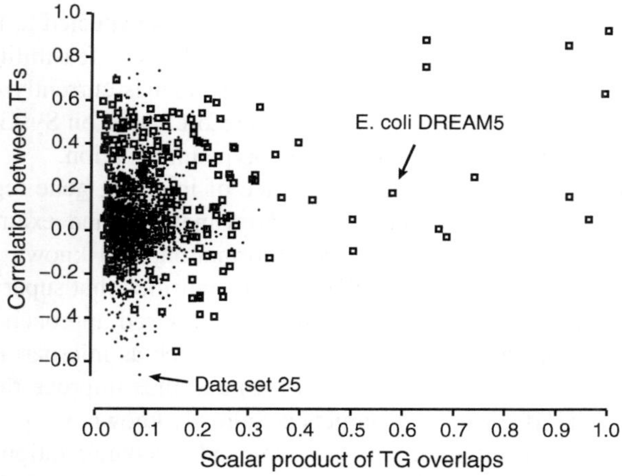

Fig. 6.8 Relationship between TF correlation and TF target genes. Each point corresponds to a pair of TFs in the SGS dataset 25 (*dots*) and an *E. coli* dataset (*squares*). The abscissa shows the overlaps between the target genes of the two transcription factors calculated as scalar product. The ordinate depicts the Pearson's correlation computed between the two TF's expression profiles. Points in the upper right corner thus refer to pairs of TFs that exhibit both a high correlation and highly overlapping sets of target genes

in SVC(*D*, *S*) only slightly increases performance (compare Fig. 6.2b), whereas it strongly increases prediction performance in case of real data (not shown).

6.4 Discussion

Several classes of methods have been successfully applied to the inference of gene regulatory networks from expression data. To deduce gene regulatory interactions, unsupervised approaches depend on some dependency between the expression profiles of TFs and targets. This procedure consequently assumes that the activity of the TF itself is regulated on the expression level. The approach might fail in case of TFs that are regulated at the protein level that is not visible in gene expression measurements. In contrast, Mordelet and Vert (2008) demonstrated that a supervised approach called SIRENE, by exploiting prior information from known gene regulatory interactions, can improve the performance of inference considerably. Variants of this approach have been proposed that integrate additional types of information (e.g., TF promoter binding predictions) to further improve performance.

Here, we examined the feasibility of network inference by supervised approaches that integrate expression data, gene regulatory interactions, and genotype information. On the one hand, a subset of genotype variations is expected to have a substantial impact on the variability of gene expression measurement. Being able to inform inference approaches with genotype information is thus expected to improve inference performance. On the other hand, a reliable analysis of feasibility of genotype supported inference requires a controlled use case that is currently only available through network simulation. This study was therefore based on SysGenSim (SGS), a well-known simulator of networks with genotype information.

In this setup, we compared the performance of inferring gene regulatory interactions by several supervised and unsupervised methods and examined how the availability of expression data, genotype information, and known gold standard interactions influences performance. We indeed confirmed that supervised methods are able to perform substantially better than unsupervised approaches. Supervised approaches can assess the importance of features (such as microarray experiments where a given TF is active, Naeem et al. 2012) and thus improve the inference of targets of a given TF if some of the targets are already known.

We first analyzed the effect of preprocessing and parameterization on the performance of supervised inference. We found that different approaches to data normalization and data representation (e.g absolute measurements vs. fold-changes) had little influence on the performance. The little effect of normalization on the performance is possibly due to the fact that the SGS data is almost normalized/standardized (personal communication with organizers). However, it is still recommended to apply normalization, especially in case of combining data from different experiments/labs. Apart from preprocessing steps, choosing different parameters and kernels for training had little effect. Given SGS simulated data, we thus observed a robust performance that was almost independent of the specific training setting.

Next, we examined whether the basic supervised setting, i.e training models on features derived from expression data, could be improved by incorporating additional features generated from the genotype information or from the known gold standard interactions. Strikingly, we found that expression data in this setup was the most informative data and integration of additional information could hardly improve performance. Apparently, from training the models on subsets of the known targets of a given regulator, genotype information could be deduced from the expression measurements so that explicitly providing that information did not lead to further performance improvements. A further explanation of this finding is that genes adjacent on the chromosome are affected similarly by recombination and thus overlap strongly in their genotypes across individuals. Consequently, in the simulated data, genotype information provides limited information on sets of contiguous regulators and genes.

We also aimed to more thoroughly exploit the known gold standard interactions for inference. The standard SIRENE approach predicts targets for a given regulator at a time and is thus not able to incorporate information on other regulators and their targets. We therefore represented the network topology of genes and their regulators explicitly as additional features to better inform the training of supervised inference models. However, this feature set extension could hardly improve performance, which is in stark contrast to our previous experience on real gene regulatory networks and datasets where a substantial increase in performance could be achieved here. In real networks, functionally related genes tend to be regulated by similar sets of TFs and, vice versa, TFs forming complexes tend to regulate similar sets of target genes. We indeed find such significant overlaps between regulator and target sets in real networks, but they are virtually absent in the SGS networks. Using more realistic network topologies could thus improve SGS simulations.

To estimate the feasibility of network inference from real systems genetics data, we examined further similarities and differences between real and SGS datasets. We first analyzed whether functionally related TFs show an increased correlation in their expression profiles. In order to compare real and artificial datasets, we quantified the functional similarity of TFs based on their sets of target genes: two TFs are assumed more related if their respective target sets exhibit higher overlaps. Functionally related TFs indeed tend to exhibit a higher correlation on the expression level (Segal et al. 2003; Remenyi et al. 2004) which can be exploited for inference in case of the real data. This is not possible for the SGS generated data, where overlaps in target sets did not lead to a higher correlation between expression profiles of the corresponding TFs. In addition to using realistic network topologies, a careful parametrization of network models would be required so that simulated and real datasets exhibit similar properties.

We also found that some other properties of expression data were qualitatively similar between real and artificial datasets. For instance, in both real and artificial data, pairs of genes regulated by the same set of TFs exhibited stronger correlations in their expression profiles than arbitrary pairs. However, this effect was noticeably more extreme in the SGS data.

As the SGS datasets and networks analyzed here represent simplified models, it will be an important future task to explore network inference assisted by genotype information in case of real systems. Real genotypes will exhibit a higher complexity with respect to both structure and effects of the genotypes. Exploiting the more complex genotype information in real organisms will be thus more challenging. First, in real systems SNPs may exhibit a much broader range of effects, e.g., they may be ineffective, kill the protein completely, or exhibit strong side effects. In comparison to SGS-generated data, only a minority of 'real' mutations will just modulate the effect of a TF on its targets, and will thus be informative for network inference. A second simplification of SGS is that each gene has a single mutation locus that can assume only two states, i.e., the genotype for a given gene is represented as a single binary value. Genes in real systems may host several mutation sites in both promoter and coding sequence that influence the function of that gene in a combinatorial manner. Third, the expression of targets genes is not only modulated by SNPs and through gene regulatory networks, but also by other mechanisms affecting the protein sequence such as alternative splicing and other mechanisms affecting the state of the biological system such as metabolic and protein networks.

These complexities may decrease the performance of inference approaches on real data and will require corresponding adjustments that might complicate their application. Exploiting this information in unsupervised inference approaches might for instance require to understand the impact of combinatorial genotypes for every single regulator or gene so that they can be modeled explicitly. In the light of these considerations it is interesting to recall our above finding that the explicit specification of the genotype might not be necessary in supervised inference approaches. The application of supervised approaches to real data might thus be more robust and straightforward than employing current unsupervised methods. Supervised approaches, on the other hand, could further be applied as feature selection approaches that evaluate which of the given features are most informative for the problem at hand. As an interesting future exploratory approach, supervised inference methods could thus determine weights of SNPs and therefore evaluate the structural basis of how SNPs influence the effect of TFs on their targets.

References

Abdulrehman D, Monteiro PT, Teixeira MC, Mira NP, Lourenco AB, dos Santos SC, Cabrito TR, Francisco AP, Madeira SC, Aires RS, Oliveira AL, SaCorreia I, Freitas AT (2011) YEASTRACT: providing a programmatic access to curated transcriptional regulatory associations in Saccharomyces cerevisiae through a web services interface. Nucleic Acids Res 39:D136–D140

Bleakley K, Biau G, Vert JP (2007) Supervised reconstruction of biological networks with local models. Bioinformatics 23:i57–i65

Chua G, Morris QD, Sopko R, Robinson MD, Ryan O, Chan ET, Frey BJ, Andrews BJ, Boone C, Hughes TR (2006) Identifying transcription factor functions and targets by phenotypic activation. Proc Natl Acad Sci USA 103:12045–12050

Ernst J, Beg QK, Kay KA, Balazsi G, Oltvai ZN, BarJoseph Z (2008) A semi-supervised method for predicting transcription factor-gene interactions in Escherichia coli. PLoS Comput Biol 4:e1000044

Ernst J, Kheradpour P, Mikkelsen TS, Shoresh N, Ward LD, Epstein CB, Zhang X, Wang L, Issner R, Coyne M, Ku M, Durham T, Kellis M, Bernstein BE (2011) Mapping and analysis of chromatin state dynamics in nine human cell types. Nature 473:43–49

Faith JJ, Hayete B, Thaden JT, Mogno I, Wierzbowski J, Cottarel G, Kasif S, Collins JJ, Gardner TS (2007) Large-scale mapping and validation of Escherichia coli transcriptional regulation from a compendium of expression profiles. PLoS Biol 5:e8

Hecker M, Lambeck S, Toepfer S, van Someren E, Guthke R (2009) Gene regulatory network inference: data integration in dynamic models—a review. Biosystems 96(1):86–103

Hu Z, Killion PJ, Iyer VR (2007) Genetic reconstruction of a functional transcriptional regulatory network. Nat Genet 39:683–687

Huynh-Thu VA, Irrthum A, Wehenkel L, Geurts P (2010) Inferring regulatory networks from expression data using tree-based methods. PLoS One 5(9). pii: e12776

Jansen RC, Nap JP (2001) Genetical genomics: the added value from segregation. Trends Genet. 17:388–391

Küffner R, Petri T, Tavakkolkhah P, Windhager L, Zimmer R (2012) Inferring gene regulatory networks by ANOVA. Bioinformatics 28:1376–1382

Lanckriet GRG, Deng M, Cristianini N, Jordan MI, Noble WS (2004) Kernel-based data fusion and its application to protein function prediction in yeast, vol 9. World Scientific Singapore, Singapore, pp 300–311

Loh PR, Tucker G, Berger B (2011) Phenotype prediction using regularized regression on genetic data in the DREAM5 systems genetics B challenge. PLoS One 6(12):e29095

Margolin AA, Nemenman I, Basso K, Wiggins C, Stolovitzky G, Califano A (2006) ARACNE: an algorithm for the reconstruction of gene regulatory networks in a mammalian cellular context. BMC Bioinformatics 7(Suppl 1):S7

Mordelet F, Vert JP (2008) SIRENE: supervised inference of regulatory networks. Bioinformatics 24:i76–i82

Mostafavi S, Ray D, Warde-Farley D, Grouios C, Morris Q (2008) GeneMANIA: a real-time multiple association network integration algorithm for predicting gene function. Genome Biol 9(Suppl 1):S4

Naeem H, Zimmer R, Tavakkolkhah P, Küffner R (2012) Rigorous assessment of gene set enrichment tests. Bioinformatics 28:1480–1486

Pinna A, Soranzo N, Hoeschele I, de la Fuente A (2011) Simulating systems genetics data with SysGenSIM. Bioinformatics. 27(17):2459–62

Reményi A, Schöler HR, Wilmanns M (2004) Combinatorial gene expression by DNA-governed alignment of proteinprotein interaction surfaces. Nat Struct Mol Biol 11:812–815

Segal E, Shapira M, Regev A, Pe'er D, Botstein D, Koller D, Friedman N (2003) Identifying regulatory modules and their condition-specific regulators from gene expression data. Nat Genet. 34(2):166–76

De Smet R, Marchal K (2010) Advantages and limitations of current network inference methods. Nat Rev Microbiol 8:717–729

Vert JP, Kanehisa M (2003) Extracting active pathways from gene expression data. Bioinformatics 19(Suppl 2):ii238–ii244

Vignes M, Vandel J, Allouche D, Ramadan-Alban N, Cierco-Ayrolles C, Schiex T, Mangin B, de Givry S (2011) Gene regulatory network reconstruction using Bayesian networks, the dantzig selector, the Lasso and their meta-analysis. PLoS One 6(12):e29165

Zheng W, Zhao H, Mancera E, Steinmetz LM, Snyder M (2010) Genetic analysis of variation in transcription factor binding in yeast. Nature 464:1187–1191

Chapter 7
Integration of Genetic Variation as External Perturbation to Reverse Engineer Regulatory Networks from Gene Expression Data

Francesco Sambo, Tiziana Sanavia and Barbara Di Camillo

Abstract In systems genetics, genetic variations can be thought as a randomized, multifactorial set of perturbations and the gene/protein expression profile of each individual as the system response to a specific set of perturbations. Current systems genetics approaches, known as genetics genomics, try to combine different types of data such as expression and genetic data both to improve the performance of reverse engineering application and to get deepest biological insights. In this chapter, we present an integrative reverse-engineering approach which exploits both genetic and expression data. The method: 1) codifies genetic-induced perturbations by a variation matrix, which represents differences of mean expressions in each gene according to the genotype of its regulators; 2) define genetic correlation blocks within the variation matrix based on the correlation between genotypes; 3) infers the correct cause-effect pairs based on local peaks of intensity in the variation matrix, since genetic correlation decreases with genetic distance from the real causal gene. Compared to other pair-wise methods typically used in reverse-engineering, the variation matrix shows good performance in terms of both area under receiver operating characteristic and area under the precision versus recall curve. However, on the StatSeq benchmark our approach is able to address a limited number of independent perturbations, due to the high genetic linkage observed in the data. The obtained results provide a basis for advanced integrative approaches able to automate the systematic interpretation of perturbation experiments exploiting the genetic-based prior knowledge.

7.1 Introduction

Cellular processes involve millions of molecules playing a coherent role in the exchange of matter, energy and information both among themselves and with the

F. Sambo · T. Sanavia · B. Di Camillo (✉)
Department of Information Engineering, University of Padova, Padua, Italy
e-mail: barbara.dicamillo@dei.unipd.it

A. de la Fuente (ed.), *Gene Network Inference*,
DOI: 10.1007/978-3-642-45161-4_7, © Springer-Verlag Berlin Heidelberg 2013

environment. These processes are regulated by proteins, whose expression is controlled by a tight network of interactions between genes, proteins, and other molecules. A major goal of systems biology is the elucidation of the complex network of interacting DNA sequences, RNAs and proteins regulating and controlling gene expression. The term “reverse engineering” indicates the set of methods useful to reconstruct a regulatory network from its observed output, obtained from either dynamic or static multiple stimulus-response experiments. Today, high-throughput technologies such as microarray and mass spectrometry can measure the expression of genes and proteins at a given instant, thus making possible, at least in principle, the reconstruction of the regulatory network from its observed output through reverse-engineering approaches. However, given the complexity of biological systems and the high number of molecules involved, reverse-engineering approaches are usually limited to very general and abstract models, where RNA expression is considered as a proxy of protein expression in controlling gene transcription.

Several reverse-engineering approaches have been proposed in the recent years to infer transcriptional regulatory networks from microarray gene expression data. Among them Boolean models (Somogyi et al. 1996; Liang et al. 1998; Shmulevich et al. 2002; Gat-Viks and Shamir 2003), models based on differential equations, (D’haeseleer et al. 1999; de la Fuente et al. 2002; Gardner et al. 2003; de Jong 2002; Sambo et al. 2012), Bayesian networks, (Friedman et al. 2000; Friedman 2004; Yu et al. 2002), and methods based on pair-wise gene expression correlations (Butte e Kohane 2000; Ferrazzi et al. 2007; Herrero et al. 2003; Basso et al. 2005; Schäfer and Strimmer 2005; Badaloni et al. 2012; Margolin et al. 2007). However, reverse-engineering methods on real datasets and realistically simulated data (Mendes et al. 2003; Di Camillo et al. 2009; Marbach et al. 2009) are known to exhibit poor performance (Bansal et al. 2007; Soranzo et al. 2007). Commonly adopted performance measures, such as Precision and Recall, rarely rise beyond 0.5 in a 0–1 scale, one of the main reasons being the impossibility to design an experiment rich enough to observe all the different states of the system responding to a randomized, multifactorial set of perturbation.

As Jansen observed, high-throughput genomics, transcriptomics, proteomics and metabolomics have the potential to identify the functional consequences of induced and natural genetic variation (Jansen 2003). In other words, genetic variations can be thought as perturbations and the gene/protein expression profile of each individual as the system response to its specific set of perturbations. Current systems genetics approaches try to combine different types of data such as expression and genetic data both to improve the performance of reverse engineering application and to get deepest biological insights. The StatSeq benchmark, presented in Chap. 1 of this book, was designed to allow training and evaluation of reverse-engineering algorithms applied to expression and genetic data. It consists of 72 datasets simulated on networks with 100, 1,000, or 5,000 genes.

In this chapter, we present our reverse-engineering approach by following the rationale of its development and motivating the design choices on the 24 datasets of size 100. The datasets of size 1,000 and 5,000 will be used to show how performance scales on networks of larger size.

7.2 Genetic Variation as Local Perturbation

The StatSeq benchmark provides genotype and gene expression data for *in-silico* populations, where each gene exhibits a single DNA polymorphism either in the promoter region (*cis-effect*) or in the coding region (*trans-effect*). Simulated polymorphisms are characterized by two possible genetic variants (coded by 0 or 1).

Both types of polymorphisms can be interpreted as multifactorial genetic perturbations and can be exploited to gain a system-level understanding of biological networks: in the cis-effect case, genetic variants affect the steady-state expression of the gene itself; in the trans-effect case, genetic variants have no effects on the expression of the gene, but the effect can be seen in the expression levels of the corresponding gene targets. In both cases, the polymorphism leads to an alteration of the expression levels of the downstream genes and can thus be treated as a local perturbation of the gene network.

Even without knowing which genetic variant is associated to the active state of the polymorphism and which type of effect (between -cis and -trans) characterizes the polymorphism, it is possible to analyze the alterations in the expression of all genes according to the variations in the genotype of each gene. Considering two genes i and j, we can thus assess if the variations of the gene j across the genotype of different subjects induce any significant change in the expression of gene i. A first attempt to quantify this kind of interaction is to consider the differences in the mean gene expression values of gene i between two groups of subjects, divided according to the two genetic variants belonging to the gene j.

We define a *variation matrix* V as an $N \times N$ matrix (with N indicating the number of genes) coding all pair-wise genetic-induced interactions. Each element (i, j) in V measures how much the genotype of j affects the expression of i as the absolute difference between the mean expression of gene i across subjects with genotype $j = 0$ and subjects with genotype $j = 1$.

The pseudocode for computing the variation matrix is as follows, with E and G representing the gene expression matrix and the binary genotype matrix, respectively:

VARIATIONMATRIX (E, G)
  E : gene expression matrix
  G : binary genotype matrix
  for each gene j
    g_0 = subjects for which j's genotype equals 0
    g_1 = subjects for which j's genotype equals 1
    for each gene i
      $m_0 = \text{mean}(E(i, g_0))$
      $m_1 = \text{mean}(E(i, g_1))$
      $V(i, j) = |m_0 - m_1|$
  return V

We computed the variation matrix for each pair of expression and genotype matrices E and G provided by the StatSeq benchmark. Figure 7.1 shows two examples of

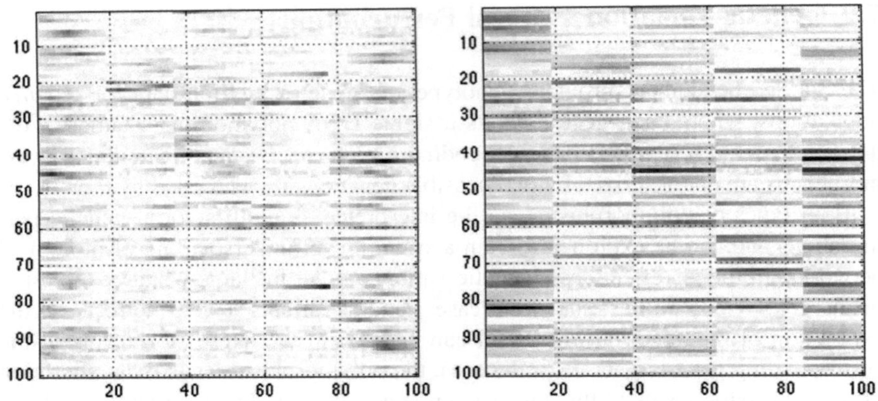

Fig. 7.1 Variation matrices for data sets 2 (*left panel*, higher average marker distance) and 6 (*right panel*, lower average marker distance) of size 100. *Darker gray* intensities correspond to greater values of absolute difference between the mean expression of gene i across subjects with genotype $j = 0$ and subjects with genotype $j = 1$

variation matrices for datasets 2 (left panel) and 6 (right panel) with intensities represented in gray scale. As a first approximation, the variation matrix can be related to the regulatory effect of gene j on gene i; however, by inspecting Fig. 7.1, one can observe horizontal lines corresponding to genes which are close to each other in the genome and show correlation in the response they induce. In other words, the simulated *genetic linkage*, i.e., the nonrandom association between portions of the genome close to each other, strongly affects the variation matrices. This effect is more evident for dataset 6 as compared to dataset 2. In fact, genetic linkage is simulated in the StatSeq benchmark with two levels of strength: lower association between markers due to a higher average marker distance (dataset 2) and higher association due to lower average distance (dataset 6). Genetic linkage implies that the genetic variation, interpreted as a local perturbation in our variation matrix analysis, cannot be considered independent for genes close to each other. The presence of genetic linkage, thus, limits the ability of identifying true causal relations directly from the variation matrix.

The strength of genetic linkage can be more clearly observed and compared by plotting the absolute correlation between the genotype of all pairs of genes (Fig. 7.3). Comparing the two correlation matrices in Fig. 7.2 with the corresponding variation matrices in Fig. 7.1, it stands out how the effect of genetic linkage is reflected also in the gene expression data, since correlated perturbations result in correlated variations of the gene expression.

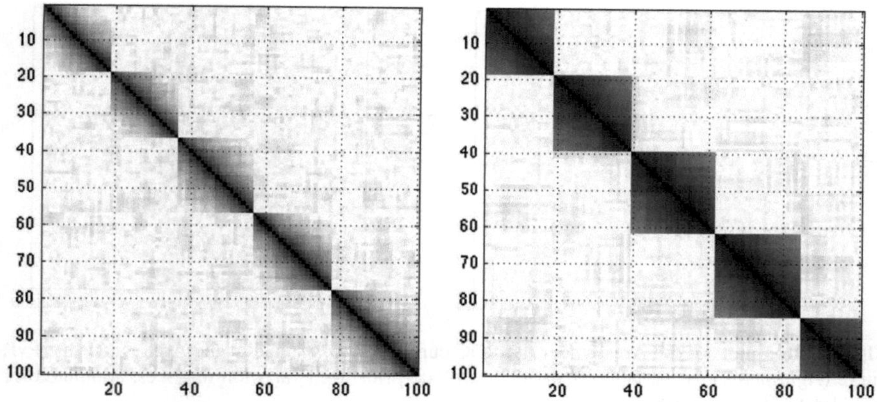

Fig. 7.2 Matrices of correlation between genotypes for data sets 2 (*left panel*, higher average marker distance) and 6 (*right panel*, lower average marker distance) of size 100. *Darker gray* intensities correspond to greater absolute values of correlation

7.3 Genetic Variation as Block-Wise External Perturbation

Considering high-valued elements of variation matrices as candidate causal relations, thus, we are deemed to identify a large number of false-positive associations due to genetic linkage. Suppose in fact to have a set of adjacent genes $j_1, j_2, \ldots, j_n$, characterized by strongly correlated genetic variations as a consequence of genetic linkage, and a gene i regulated just by one or few of the j_k: if we infer regulatory relations by the strength of the effect of genetic variations, i will appear to be regulated by the entire set $j_1, j_2, \ldots, j_n$.

Despite the confounding effect of genetic linkage, however, the correct cause–effect pair will tend to be associated, with high probability, to a local peak of intensity in the corresponding row of the variation matrix: since genetic correlation decreases with genetic distance from the real causal gene, so will, in general, the intensity in the variation matrix. Maximal elements of each correlated block in each row thus have higher probability of corresponding to real regulatory relations.

Stemming from this idea, we developed an algorithm for prioritizing the maxima in each row and each correlation block (i.e., each simulated chromosome) of the variation matrix. The pseudocode of the algorithm is as follows:

BLOCK-PRIORITIZE(A, G)

A : input matrix

G : binary genotype matrix

θ = threshold of absolute pair-wise correlation (0.2, in our experiments)

C = PAIR-WISE CORRELATION (G)

Identify the borders between chromosomes, as the pairs $(i, i+1)$ of consecutive genes for which $|C(i, i+1)| < \theta$

Chromosomes = list of vectors, one for each chromosome *ch*, each containing the indices of the genes in *ch*

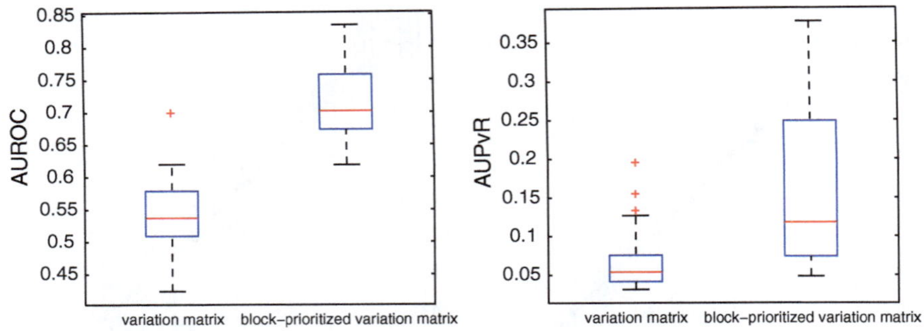

Fig. 7.3 Box plots of the Area Under the ROC curve (*left panel*) and of the Area Under the PvR curve (*right panel*) for variation matrices and block-prioritized variation matrices, computed for the 24 datasets of size 100

$M = \max(A)$
for each gene i and chromosome ch
$j = \text{argmax}(A(i, Chromosomes\{ch\}))$
$A(i, j) + = M$
return A

We name this procedure *block-prioritization* and apply it to the variation matrix. If the elements of the matrix are interpreted as levels of confidence on the corresponding cause–effect relations, block-prioritization assigns the highest confidence to the identified maxima. This way, we impose to each gene i a major regulator j in each chromosome ch. *Block-prioritization* was designed ad hoc for the StatSeq benchmark, to cope with the limited number of independent perturbations due to the high genetic linkage observed in the simulated data. However, this way we limit the number of false positives at the price of a large number of false negatives. The threshold θ of absolute pairwise correlation (*Block-prioritization*, line 3) was fixed to 0.2 with the only purpose of identifying the borders between chromosomes on the simulated data. On real data, where genetic linkage pattern is more blurred, we do not expect the need for using the *block-prioritization* procedure.

We tested the performance in reverse engineering of both variation matrices and block-prioritized variation matrices, on the 24 simulated datasets of size 100, in terms of Area under the Receiver Operating Characteristic (AUROC) and Area under the Precision versus Recall Curve (AUPvR). As it can be seen from Fig. 7.3, the application of block-prioritization significantly increases the performance in terms of both AUROC (left panel) and AUPvR (right panel) as compared to the variation matrix representation.

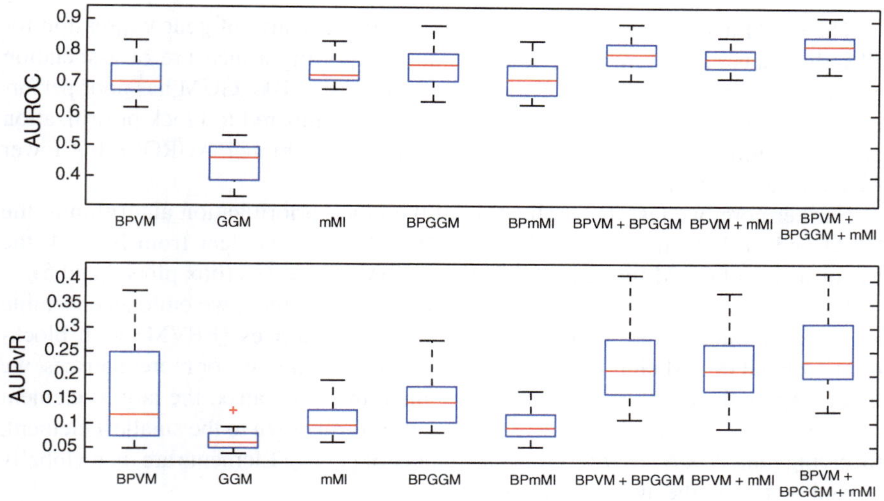

Fig. 7.4 Box plots of the Area Under the ROC curve *(top)* and of the Area Under the PvR curve *(bottom)* obtained, on the 24 datasets of size 100, by rank average of block-prioritized variation matrices (BPVM) with different combinations of matrices. From *left* to *right*: without other matrices, with GGM, with mMI, with both GGM and mMI, with block-prioritized GGM, with block-prioritized mMI and with GGM and mMI, both block-prioritized

7.4 Improving Performance with Pair-Wise Methods

Given the proven effectiveness of Graphical Gaussian Models (GGM) and minimum conditional Mutual Information (mMI) in reverse engineering large-scale gene regulatory networks (Soranzo et al. 2007), we explored how to best exploit the two procedures for further improving the performance of our algorithm.

Graphical Gaussian Models assess the strength of a relation between two genes as their partial correlation conditioned on all the other genes. We adopted the approach described in (Schäfer and Strimmer 2005) and implemented in the R package GeneNet, version 1.2.5 (http://strimmerlab.org/software/genenet/): the partial correlation is estimated through the Moore-Penrose pseudo-inverse of the correlation matrix and the estimate is stabilized with Bootstrap resampling.

Mutual Information is another measure of dependency between random variables, arising from information theory, which has been proposed in the literature for assessing the strength of a relation between two genes (Butte and Kohane 2000; Daub et al. 2004). Following the approach described in Soranzo et al. (2007) and implemented in the C++/MATLAB package NetRec (http://people.sissa.it/~altafini/papers/SoBiAl07/), we considered for each pair of genes the minimal Mutual Information conditioned on each of the other genes. Mutual Information is computed with the B-spline algorithm of (Daub et al. 2004).

The GGM and mMI algorithms process the entire matrix of gene expression for all individuals and return a symmetric matrix, reporting a measure of association between all gene pairs. Tested on the 24 datasets of size 100, GGM has poor performance in terms of both AUROCs and AUPvRs as compared to block prioritization of the variation matrix (BPVM), while mMI exhibits higher AUROCs but lower AUPvRs, as shown in Fig. 7.4 (box plots 1–3).

Furthermore, we test the application of the block-prioritization algorithm to the GGM and mMI matrices (BPGGM and BPmMI): as it is clear from Fig. 7.4, the performance of GGM increases while that of mMI decreases (box plots 4 and 5).

Finally, to exploit the information gained by each method, we build an ensemble network by integrating block-prioritized variation matrices (BPVM) with block-prioritized GGM (BPGGM) and simple mMI. To integrate two or more matrices, we first compute the absolute rank of each element in each matrix: the largest element receives rank 1, the second largest rank 2, and so forth down to the smallest element, receiving rank $N \times N = N^2$ (if N is the number of genes). Elements are then globally ranked by taking the average rank across matrices.

As it is clear from Fig. 7.4, combining the BPVM matrix with either the mMI matrix or the block-prioritized GGM matrix leads to a significant increase in AUROC and AUPvR (box plots 5 and 6). The effect is even stronger when the three matrices are combined (box plot 7).

Given the experimental results on networks of size 100, our final reverse-engineering algorithm is thus as follows:

```
REVERSEENGINEERING(E, G)
    E : GENE EXPRESSION MATRIX
    G : BINARY GENOTYPE MATRIX
    VM = VARIATIONMATRIX(E, G)
    GGM = GRAPHICALGAUSSIANMODELS(E)
    MMI = MINIMUMCONDITIONALMI(E)
    BPVM = BLOCK-PRIORITIZE(VM, G)
    BPGGM = BLOCK-PRIORITIZE(GGM, G)
    R = MEAN(RANK(BPVM), RANK(BPGGM), RANK(MMI))
    return R
```

7.5 Experimental Results

The complete reverse-engineering algorithm was tested on datasets of size 1,000 and 5,000 from the StatSeq benchmark, to assess how performance scales with the number of genes. Comparisons throughout the section are carried out with the two-tailed Wilcoxon sign-rank test and differences are considered significant for p-values below 0.05.

We were not able to compute mMI for networks of size 5,000 in a reasonable time: the complexity of the mMI computation is cubic in the number of genes and

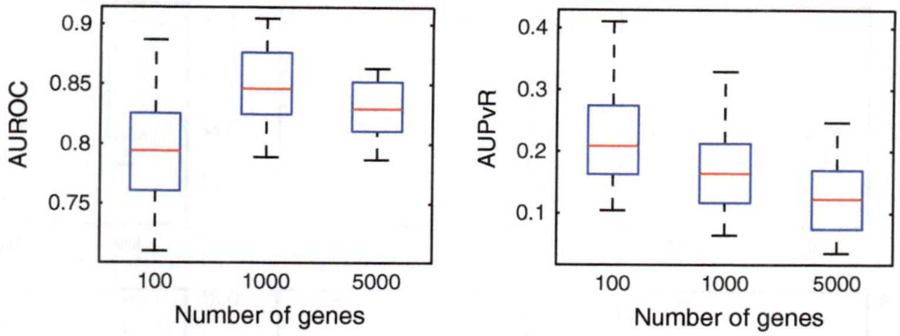

Fig. 7.5 Box plots of the Area Under the ROC curve (*left*) and of the Area Under the PvR curve (*right*) obtained on 24 datasets of size 100, 1,000, and 5,000. For datasets of size 100 and 1,000, BPVM, BPGGM, and mMI results were integrated, whereas for dataset of size 5,000, only BPVM and BPGGM results were integrated

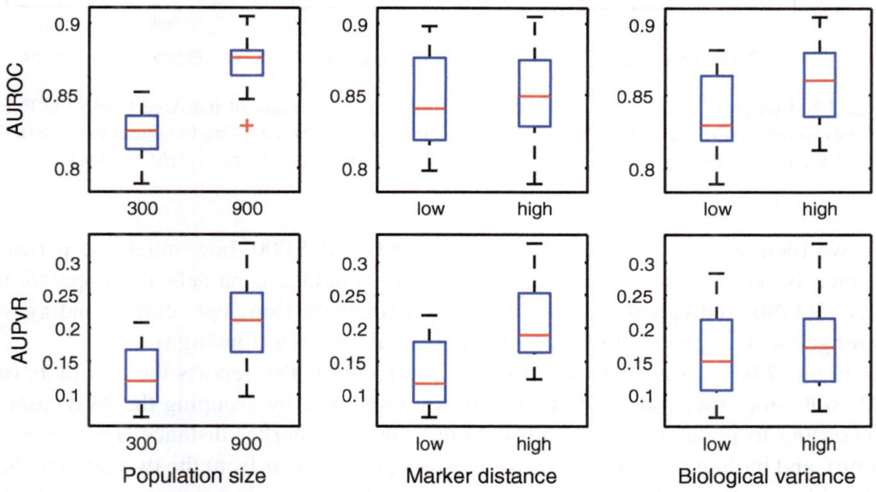

Fig. 7.6 Box plots of the Area Under the ROC curve (*top row*) and of the Area Under the PvR curve (*bottom row*) obtained by grouping the 24 datasets of size 1,000 according to population size (*left column*), average marker distance (*middle column*) and biological variance (*right column*)

the estimated time for completion of a single dataset is between 10 and 20 days. For this reason, we integrated BPVM, BPGGM, and mMI for datasets of size 1,000 and just the BPVM and BPGGM for datasets of size 5,000.

We report in Fig. 7.5 the box plots of AUROC and AUPvR for the three dataset sizes 100, 1,000, and 5,000. As it is clear from the figure, AUROC and AUPvR exhibit a different behavior: while AUPvR monotonically decreases with size (all p-values $\leq 3 \times 10^{-3}$), AUROC increases for networks of size 1,000 (p-value 2×10^{-5}) and then decreases again for size 5,000 (p-value 6×10^{-3}). This last drop in AUROC might be imputed to the unavailability of the mMI matrix for datasets of size 5,000.

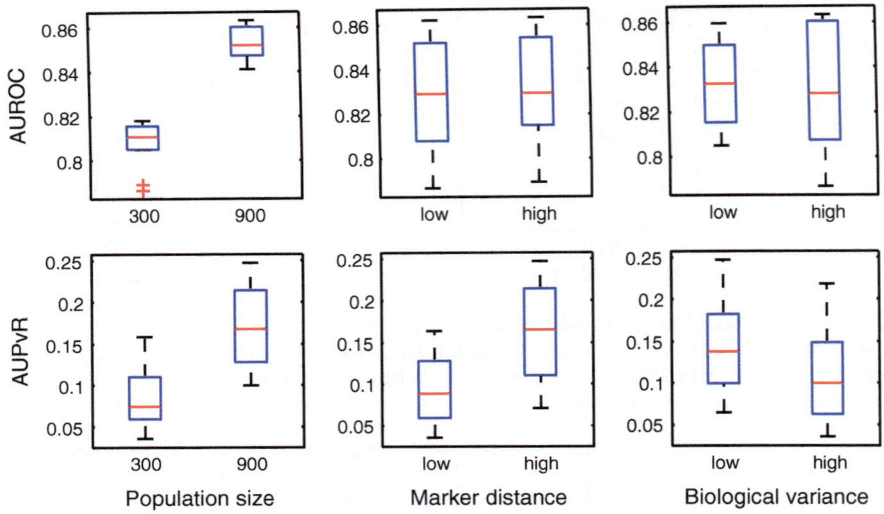

Fig. 7.7 Box plots of the Area Under the ROC curve (*top row*) and of the Area Under the PvR curve (*bottom row*) obtained by grouping 24 datasets of size 5,000 according to population size (*left column*), average marker distance (*middle column*) and biological variance (*right column*)

We then assessed, for datasets of size 1,000 and 5,000, how much the performance is sensitive to three parameters of the simulation, namely the population size (300/900 individuals), the average marker distance (low/high, corresponding to stronger/weaker genetic linkage), and biological variance (low/high).

Figure 7.6 (for size 1,000) and Fig. 7.7 (for size 5,000) reports the box plots of AUROC (top row) and AUPvR (bottom row) obtained by grouping the 24 datasets according to population size (left column), average marker distance (middle column), and biological variance (right column). As it is clear from the two figures, the algorithm significantly benefits from increasing the population size in terms of both AUROC and AUPvR, for both dataset sizes (all p-values $< 5 \times 10^{-5}$).

A higher average marker distance positively affects AUPvR for both datasets sizes (both p-values $< 5 \times 10^{-5}$), whereas it exhibits a weaker but significant effect on AUROC for size 5,000 (p-value 0.027) and no significant effect for size 1,000 (p-value 0.176).

Finally, the effect of biological variance on performance seems to have an opposite trend for sizes 1,000 and 5,000: in the former case, an increase in biological variance appears beneficial, though only significant for AUROC (p-value 5×10^{-5}) and not for AUPvR (p-value 0.077); in the latter case, increasing biological variance significantly deteriorates AUPvR (p-value 5×10^{-5}) but not AUROC (p-value 0.569).

7.6 Conclusions

In this chapter we presented a simple approach, integrating genetic and expression data and different reverse-engineering methods to reconstruct the regulatory network underlying the observed gene expression. Our method was designed ad hoc on the StatSeq benchmark, based on the results obtained on the 24 datasets of 100 genes.

The original plan, however, was to apply the variation matrix algorithm to distinguish between -cis and -trans effects and then to infer regulatory relations with SPQR (Systematic Perturbation—Qualitative Reasoning), a novel Qualitative Reasoning approach to automate the interpretation of the results of systematic perturbation experiments (Badaloni et al. 2012). SPQR exploits a set of IF-THEN rules to infer causal relations between the variables, analyzing the patterns of propagation of the perturbation signals through the biological network, and is specifically designed to minimize the rate of false positives among the inferred relations.

This latter strategy was not performing well on the StatSeq benchmark, due to the high genetic linkage observed in the data. In fact, the systematic perturbation of the components of a biological system has been proven among the most informative experimental setups for the identification of causal relations between the components. However, the high genetic linkage observed in the simulated data limited the actual number of independent perturbations. For example, in the dataset of 100 genes, we observed basically five independent stimuli corresponding to the five blocks of highly correlated genotypes shown in Fig. 7.2.

In real data, the distinction between -cis and -trans effect of a genetic variant can be reasonably inferred from the variant position; moreover, correlation between genetic markers is expected to be lower than the one observed here, especially if one is trying to reconstruct regulatory networks involving genes with high distance in the genome. Our future plan is thus to test the application of SPQR on real eQTL data. On the other hand, the block-prioritized approach developed here proved to be a valid representation of the genetic-induced interactions, robust to the indeterminateness intrinsically introduced by high genetic linkage, and it can be easily integrated in SPQR to optimize its performance when analyzing regions of high genetic correlation.

References

Badaloni S, Di Camillo B, Sambo F (2012) Qualitative reasoning for biological network inference from systematic perturbation experiments. IEEE/ACM Trans Comput Biol Bioinform 9(5):1482–1491

Bansal M, Belcastro V, Ambesi-Impiombato A, di Bernardo D (2007) How to infer gene networks from expression profiles. Mol Syst Biol 3:78

Basso K, Margolin AA, Stolovitzky G, Klein U, Dalla-Favera R, Califano A (2005) Reverse engineering of regulatory networks in human B cells. Nat Genet 37(4):382–390

Butte AJ, Kohane IS (2000) Mutual information relevance networks: functional genomic clustering using pairwise entropy measurements. Pac Symp Biocomput 4:418–429

Daub CO, Steuer R, Selbig J, Kloska S (2004) Estimating mutual information using B-spline functions—an improved similarity measure for analysing gene expression data. BMC Bioinform 5:118

Di Camillo B, Toffolo G, Cobelli C (2009) A gene network simulator to assess reverse engineering algorithms. Ann N Y Acad Sci 1158:125–142

D'haeseleer P, Wen X, Fuhrman S, Somogyi R (1999) Linear modeling of mRNA expression levels during CNS development and injury. Pac Symp Biocomput 4:41–52

Ferrazzi F, Sebastiani P, Ramoni MF, Bellazzi R (2007) Bayesian approaches to reverse engineer cellular systems: a simulation study on nonlinear Gaussian networks. BMC Bioinform 8(Suppl 5):S2

Friedman N, Linial M, Nachman I, Pe'er D (2000) Using Bayesian networks to analyze expression data. J Comput Biol 7:601–620

Friedman N (2004) Inferring cellular networks using probabilistic graphical models. Science 303:799–805

de la Fuente A, Brazhnik P, Mendes P (2002) Linking the genes: inferring quantitative gene networks from microarray data. Trends Genet 18:395–398

Gardner TS, di Bernardo D, Lorenz D, Collins JJ (2003) Inferring genetic networks and identifying compound mode of action via expression profiling. Science 301:102–105

Gat-Viks I, Shamir R (2003) Chain functions and scoring functions in genetic networks. Bioinformatics 19(Suppl 1):i108–i117

Herrero J, Diaz-Uriarte R, Dopazo J (2003) An approach to inferring transcriptional regulation among genes from large-scale expression data. Comp Funct Genom 4:148–154

Jansen RC (2003) Studying complex biological systems using multifactorial perturbation. Nat Rev Genet 4:145–151

de Jong H (2002) Modeling and simulation of genetic regulatory systems: a literature review. J Comput Biol 9:67–103

Liang S, Fuhrman S, Somogyi R (1998) REVEAL, a general reverse engineering algorithm for inference of genetic network architectures. Pac Symp Biocomput 3:18–29

Marbach D, Schaffter T, Mattiussi C, Floreano D (2009) Generating realistic in silico gene networks for performance assessment of reverse engineering methods. J Comput Biol 16(2):229–239

Margolin AA, Nemenman I, Basso K, Wiggins C, Stolovitzky G, Califano A (2007) ARACNE: an algorithm for the reconstruction of gene regulatory networks in a mammalian cellular context. BMC Bioinform 7(Suppl 1):S7

Mendes P, Sha W, Ye K (2003) Artificial gene networks for objective comparison of analysis algorithms. Bioinformatics 19(Suppl 2): ii122-ii129

Sambo F, Montes de Oca MA (2012) MORE: mixed optimization for reverse engineering-an application to modeling biological networks response via sparse systems of nonlinear differential equations. IEEE/ACM Trans Comput Biol Bioinform 9(5):1459–1471

Schäfer J, Strimmer K (2005) An empirical Bayes approach to inferring large-scale gene association networks. Bioinformatics 21(6):754–764

Shmulevich I, Dougherty ER, Zhang W (2002) From Boolean to probabilistic Boolean networks as models of genetic regulatory networks. Proc IEEE 90(11):1778–1792

Somogyi R, Fuhrman S, Askenazi M, Wuensche A (1996) The gene expression matrix: towards the extraction of genetic network architectures. Nonlin Anal Theory Meth Appl 30(3):1815–1824

Soranzo N, Bianconi G, Altafini C (2007) Comparing association network algorithms for reverse engineering of large-scale gene regulatory networks: synthetic versus real data. Bioinformatics 23(13):1640–1647

Yu J, Smith VA, Wang P, et al. (2002) Using Bayesian network inference algorithms to recover molecular genetic regulatory networks. Proc Intern Conf Syst Biol 99:12783–12788

Chapter 8
Using Simulated Data to Evaluate Bayesian Network Approach for Integrating Diverse Data

Luan Lin and Jun Zhu

Abstract Large-scale high-dimensional omics data sets have been generate to survey complex biological systems. However, it is a challenge how to integrate multiple dimensions of biological data to biological causal networks where comprehensive knowledge can be derived in contexts. We developed a RIMBANet (Reconstructing Integrative Molecular Bayesian Networks) method to integrate diverse biological data. In this chapter, we disseminate results of applying our RIMBANet method on a series of simulated datasets. Two sets of networks are inferred with or without integrating genetic markers with gene expression data. We show that integration of genetic data into network reconstruction using RIMBANet approach improves network construction accuracy. Furthermore, false-positive links in reconstructed networks are not randomly distributed. More than 80 % of them connect nodes that are indirect neighbors.

8.1 Introduction

Biological systems are complex. Cells employ multiple levels of regulation that enable them to respond to genetic and environmental perturbations. Advances in high-throughput technologies in biology in the past few years have created a plethora

L. Lin · J. Zhu
Department of Genetics and Genomic Sciences, New York, NY, USA

L. Lin · J. Zhu
Icahn Institute of Genomics and Multiscale Biology, New York, NY, USA

J. Zhu (✉)
The Tisch Cancer Institute, New York, NY, USA
e-mail: jun.zhu@mssm.edu

J. Zhu
Graduate School of Biomedical Sciences, Icahn School of Medicine at Mount Sinai, One Gustave L. Levy Place, PO Box 1498, New York, NY 10029, USA

A. de la Fuente (ed.), *Gene Network Inference*,
DOI: 10.1007/978-3-642-45161-4_8,

of data measuring different levels of cell regulations, such as genetics, genomics, transcriptomics, proteomics, metabolomics, glycomics, etc. Despite the complexity of biological systems, research in the context of large-scale high-dimensional omics data has tended to focus on single data dimensions. While we achieve some understanding in this way, progress is limited because none of the dimensions on their own provide a comprehensive context in which results can be fully interpreted. For example, genome-wide association studies (GWAS) have identified many hundreds of highly replicated loci associated with disease, but our understanding of underlying molecular mechanisms is still limited because the genetic loci do not necessarily inform on the gene affected, on how gene function is altered, or more generally, how the biological processes involving a given gene are altered (Altshuler 2008; Chen 2008; Emilsson 2008; Witte 2010). It is apparent that if different biological data dimensions could be integrated into biological networks, we would achieve a more complete understanding of biological systems (Chen 2008; Emilsson 2008; Hsu 2010; Schadt 2008; Zhong 2010).

Many methods for integrating high-dimensional data into networks are emerging. For example, we recently developed methods that simultaneously integrate DNA variation and RNA expression data generated in a population context to identify coherent modules of interconnected gene expression traits driven by common genetic factors (Chen 2008; Zhang et al. 2010). In addition, many groups began incorporating a time dimension in the context of high-dimensional molecular profiling data to elucidate how networks can transform over time (Leonardson 2010; Zhu 2010).

Many computational methods have been developed to utilize high-throughput data to generate comprehensive networks for elucidating biological processes and disease phenotypes. However, how to assess different computational methods for biological network construction with respect to aiding our understanding of cell regulation, elucidating complex biological processes, and generating new insights of molecular mechanisms underlying disease phenotypes, remains a difficult task. Ideally, series of prospective validation experiments are needed to systematically evaluate each computational method. However, it is costly and time-consuming. To save time and cost of experimental validations, many simulated datasets have been generated to evaluate different network reconstruction methods (Zhu 2007; Vignes 2011). Several drawbacks of the approach are (1) the simulation scheme may not truly reflect biological systems; (2) methods that share similarity of computational models underlying simulation datasets are likely to perform better. Alternatively, many hybrid schemes are created for assessing computational methods, such as DREAM (Marbach 2012), where experimental data is generated and methods are evaluated for how well the true biological regulations underlying the experimental data can be recovered. However, the evaluation results not only depend on computational methods, but also on how well participating scientists intellectually understand these biological systems. Here we disseminate results of applying our RIMBANet method (Zhu 2012, 2008) on a series of simulated datasets (Vignes 2011).

8.2 Methods and Results

8.2.1 Bayesian Network

Many methods have been proposed or developed for constructing molecular networks, some of which can only deal with linear relationships, and some of which are not easy to incorporate prior information. We have developed RIMBANet (**R**econstructing **I**ntegrative **M** olecular **Ba**yesian **Net**works) (Zhu 2012, 2008), a tool based on Bayesian framework which permits both linear and nonlinear regulation identification and which facilitates the incorporation of prior knowledge. Bayesian networks are directed acyclic graphs in which the edges of the graph are defined by conditional probabilities that characterize the distribution of states of each node given the state of its parents (Pearl 1988). The network topology defines a partitioned joint probability distribution over all nodes in a network, such that the probability distribution of states of a node depends only on the states of its parent nodes: formally, a joint probability distribution $p(X)$ on a set of nodes X can be decomposed as $p(X) = \prod_i p(X^i|\mathrm{Pa}(X^i))$, where $\mathrm{Pa}(X^i)$ represents the parent set of X^i. In our networks, each node represents a quantitative trait which can be the expression level of a gene or the abundance of a protein or metabolite, the genotype at a locus, the DNA copy variation at locus, etc. These conditional probabilities reflect not only relationships between nodes, but also the stochastic nature of these relationships, as well as noise in the data used to reconstruct the network. The simulated datasets under study consist of genetic markers and gene expression traits. Genotype data is discrete. Expression traits can be taken directly as continuous variables or be discretized as discrete variables. In the hybrid model with both continuous and discrete variables, discrete variables can be parent nodes of continuous nodes, but cannot be children of continuous nodes. In the result, we present only discrete models. Hybrid models require significantly more computational time than discrete models. The performance gain by applying hybrid models is neglectable in a few cases that we tested.

8.2.2 Learning Bayesian Network Structures

Bayes formula allows us to determine the maximum of a posterior probability $\mathrm{P}(\boldsymbol{M}|\boldsymbol{D})$ of a network model $\boldsymbol{M}$ given observed data $\boldsymbol{D}$ as a function of $\mathrm{P}(\boldsymbol{M})$—our prior belief that the model is correct and $\mathrm{P}(\boldsymbol{D}|\boldsymbol{M})$—the maximum marginal likelihood of the observed data given the model: $\mathrm{P}(\boldsymbol{M}|\boldsymbol{D}) \sim \mathrm{P}(\boldsymbol{D}|\boldsymbol{M}) * \mathrm{P}(\boldsymbol{M})$. The constant in the equation depends only on the data and is fixed given a dataset. The number of possible network structures grows super-exponentially with the number of nodes, so an exhaustive search of all possible structures to find the one best supported by the data is not feasible, as soon as the number of nodes exceeds ten or so. We implemented a local search algorithm (Friedman et al. 2000), shown in Fig. 8.1.

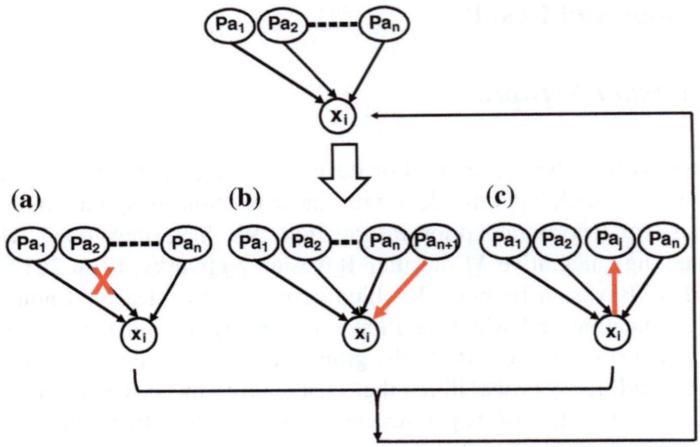

Fig. 8.1 A flow diagram illustrating local search algorithm for learning Bayesian network structure

8.2.2.1 Local Search Algorithm

The *local search* algorithm is a local model selection method, which starts with selecting a random node X_i with n parents $\mathbf{Pa} = \{Pa_1, Pa_2 \ldots, Pa_n\}$. Then, we find local structure variants that are possible for the node X_i by (a) removing a parent node; (b) adding a new parent node from a pool of potential parents; (c) reversing a parent–child relationship if the resulted structure does not violate the directed acyclic graph requirement. How to evaluate a modified structure and how to choose a criterion for accepting or not a resulted structure variation will be discussed in Sect. 8.2.2.3. After accepting or rejecting a proposed structural change at the node X_i, we randomly select the next node X_i, and iterate the process until the global structure is stable (no structure modification is accepted in $3 \times n$ trials).

8.2.2.2 Relation to Reversible Jumping in MCMC

The local search algorithm is closely related to the reversible jumping in Markov chain Monte Carlo (MCMC). The original model and proposed models have different dimensions (different numbers of parameters to estimate from data). For example in Fig. 8.1, the proposed mode1 Fig. 8.1a has less parameters to estimate than the original model. A transition function for jumping between models with different dimensions in reversible jump MCMC reflects the principle of penalizing model complexity in the model selection as discussed in Sect. 8.2.2.3.

8.2.2.3 Criteria for Model Structure Selection

A network model $\boldsymbol{M}$ consists of both a network structure and a set of parameters $\boldsymbol{\theta}$ describing conditional probabilities. Here, we refer to a network model $\boldsymbol{M}$ mainly as a network structure because the set of parameters $\boldsymbol{\theta}$ that give rise to the maximum likelihood given a network structure are used in our calculation. Given a given network model $\boldsymbol{M}$, we can evaluate it by its posterior probability $\log\mathrm{P}(\boldsymbol{M}|\boldsymbol{D}) \sim \log\mathrm{P}(\boldsymbol{D}|\boldsymbol{M}) + \log\mathrm{P}(\boldsymbol{M})$. For simplicity, we assume all structures are equal likely. Schwarz (Schwarz 1978) developed an approximation for the marginal likelihood as Bayesian Information Criterion (BIC) as the following:

$\log\mathrm{P}(\boldsymbol{D}|\boldsymbol{M}) \approx logP(D|M,\hat{\theta}) - 0.5d * logN$, where $P(D|M,\hat{\theta})$ is the maximum likelihood given the model $\boldsymbol{M}$, d is the number of parameters for the model $\boldsymbol{M}$, and N is the number of samples. Intuitively, BIC score penalizes complex models. To avoid being trapped in local maximum, we apply simulated annealing so that a structure with smaller posterior probability can also be accepted according to a rejection function which takes into account the difference in posterior probabilities of current and proposed structures and annealing temperature (reverse proportional to the number of proceed moves).

8.2.2.4 Averaged Model Selection

Searching optimal BN structures given a dataset is an NP-hard problem. We employed Monte Carlo Markov Chain (MCMC) (Madigan 1995) simulation to identify potentially hundreds of thousands of different plausible networks as described above. As the method is stochastic and the potential structure space is huge, the resulting structure will very likely be different for each run. In our process, 1,000 BNs were reconstructed using different random seeds to start the stochastic reconstruction process. From the resulting set of 1,000 networks generated by this process, edges that appeared in greater than 30 % of the networks were used to define a consensus network. A 30 % cutoff threshold for edge inclusion was based on our simulation study (Zhu 2007), where a 30 % cutoff yields the best tradeoff between recall rate and precision. In practice, cutoffs around 30 % (which have the lowest number of links in the histogram, shown in Fig. 8.2) results robustly similar structures. The histogram of edges shown in Fig. 8.2 suggests that there are two types of edges: highly recurrent ones that may represent high confident signals and sporadic recurrent ones that may represent noise. The goal of this averaging step is to keep as many high confident edges as possible and remove as many as low confident edges. In addition, a small variation of the cutoff value will result in a small number of edges, inclusion or exclusion. A cutoff around 30 % recurrence clearly separates edges into high and low confident groups. It is also the minimum of the histogram so that 30 % is chosen as the cutoff value. The consensus network resulting from the averaging process may not be a BN (a directed acyclic graph). To ensure the consensus network structure is a directed acyclic graph, edges in this consensus network were trimmed if and only if (1) the edge was involved in a loop, and (2) the edge was the most weakly supported

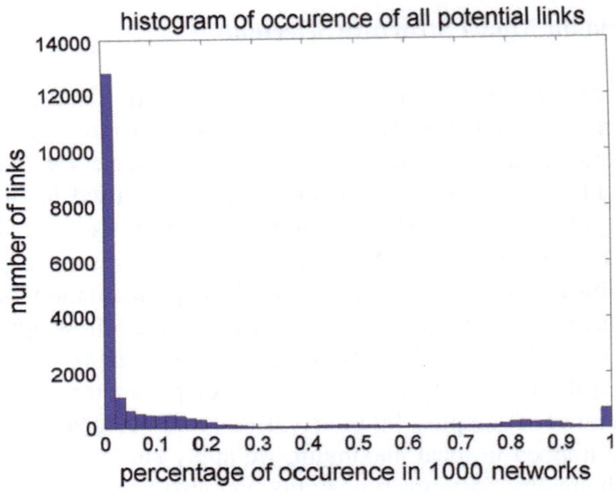

Fig. 8.2 A histogram of consensus links among 1000 independently reconstructed Bayesian network structures

of all edges making up the loop. In a case that A–> B and B–> A are equal likely, one directed edge is randomly chosen to keep.

8.2.3 Bayesian Network as a Causal Structure

Even though edges in Bayesian networks are directed, we cannot infer causal relationships from the structure directly in general. For example, in a network with two nodes, X^1 and X^2, the two models $X^1 \rightarrow X^2$ and $X^2 \rightarrow X^1$ have equal probability distributions as $p(X^1, X^2) = p(X^2|X^1)p(X^1) = p(X^1|X^2)p(X^2)$. Thus, by data itself, we cannot infer whether X^1 is causal to X^2, or vice versa. In a more general case, a network with three nodes, X^1,X^2, and X^3, there are multiple groups of structures that are mathematically equivalent. For example, the following three different models (shown in Fig. 8.3), M1 : $X^1 \rightarrow X^2, X^2 \rightarrow X^3$, M2 : $X^2 \rightarrow X^1, X^2 \rightarrow X^3$, and M3 : $X^2 \rightarrow X^1, X^3 \rightarrow X^2$, are Markov equivalent (which means that they all encode for the same conditional independent relationships). In the above case, all three structures encode the same conditional independent relationship, $X^1 \not\perp X^3|X^2$, X^1 and X^3 are independent conditioning on X^2, and they are mathematically equal

$$\begin{aligned} p(X) &= p(\text{M1}|D) = p(X^2|X^1)p(X^1)p(X^3|X^2) \\ &= p(\text{M2}|D) = p(X^1|X^2)p(X^2)p(X^3|X^2). \\ &= p(\text{M3}|D) = p(X^2|X^3)p(X^3)p(X^1|X^2) \end{aligned}$$

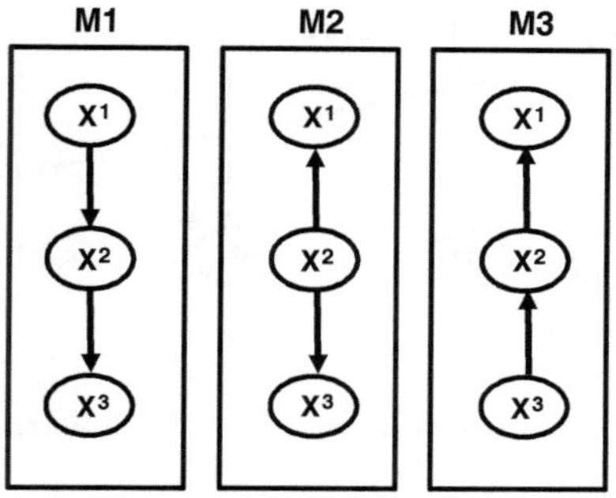

Fig. 8.3 A set of Markov equivalent models (M1-3)

Thus, we cannot infer whether X^1 is causal to X^2 or vice versa from these types of structures. However, there is a class of structures, V-shape structure (e.g., Mv : $X^1 \rightarrow X^2, X^3 \rightarrow X^2$), which has no Markov equivalent structure. In this case, we can infer causal relationships. There are more parameters to estimate in the Mv model than M1, M2, or M3, which means a large penalty in BIC score for the Mv model. In practice, a large sample size is needed to differentiate the Mv model from the M1, M2, or M3 models.

8.2.3.1 Incorporating Genetic Data as a Structure Prior

In general, Bayesian networks can only be determined by their Markov equivalent structures, so that it is often not possible to determine the causal direction of a link between two nodes even though Bayesian networks are directed graphs. In a biological system, DNA variations can cause changes in gene expression levels or protein abundance, but not the other way around. Taking this into account, the Bayesian network reconstruction algorithm—RIMBNet that we developed can incorporate genetic data to break the symmetry among nodes in the network that lead to Markov equivalent structures, thereby providing a way to infer causal directions in the network in an unambiguous fashion (Zhu 2004). The genetic priors can be derived from multiple basic sources (Zhu 2007). In the application for these simulated data, we incorporate genetic marker data as additional nodes in the Bayesian network reconstruction process, shown in Fig. 8.4. It is only possible that genetic locus regulates gene expression nodes (information can only flow from genetic variation to transcriptional changes), but gene expression node cannot regulate genetic mark nodes.

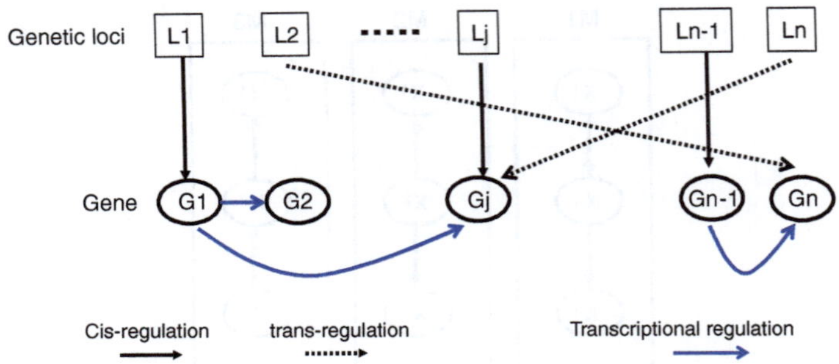

Fig. 8.4 A network structure model with two types of nodes for genetic markers and gene expression

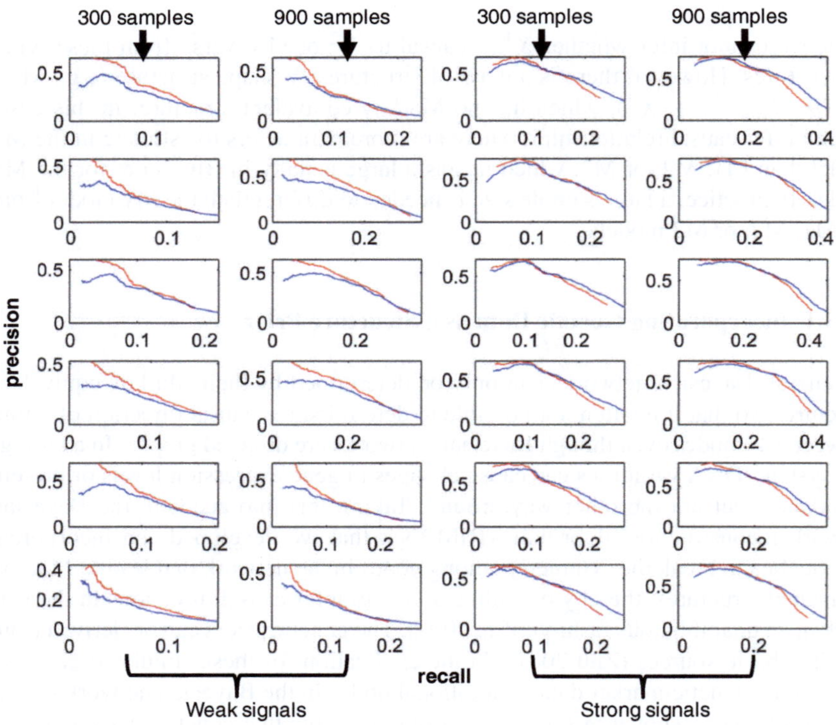

Fig. 8.5 Precision-recall plots for networks reconstructed for 24 sets of data with 1,000 nodes. The red curves are for structures reconstructed with integration of genetic marker data. The blue curves are for structures reconstructed without considering genetic marker data

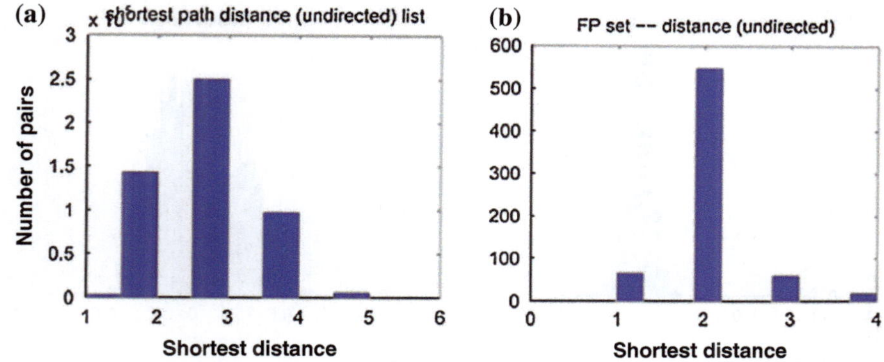

Fig. 8.6 Shortest path distributions for random pairs (**a**) and false positive pairs predicted in the reconstructed networks (**b**). Most false positive pairs are closely linked in the true networks

8.2.4 Comparing Reconstructed Structures with True Network Structures Underlying Simulated Data

We applied RIMBAnet network reconstruction algorithm to a set of simulated data for networks with 1,000 genes. We inferred two sets of networks using genetic marker data or not. Each reconstructed network is the averaged network structure based on 1,000 independent runs (see Sect. 8.2.2.4 for details). Network edges that occur in more than 30 % 1,000 individual runs are included in the final reconstructed network. In each constructed network, network edges are rank ordered by their recurrences in 1,000 individual runs. Then, we compared them to the true corresponding network structures used for data simulation. We calculated precision = true positive/total predictions, and recall = true positive links/total true links, for each reconstructed structure. Results are shown in Fig. 8.5, where red lines are for structures with genetic marker data integrated and blue lines are for structures without considering genetic marker data. When comparing lines in columns 1 and 3 with corresponding lines in columns 2 and 4, it is clear that the sample size is an important factor affecting network reconstruction accuracy. Comparing network reconstructions with the same underlying true network structure, the same number of samples but with different signal strength in the simulated data (e.g., comparing blue line in the first column in the first row with the blue line in the third column in the first row), it is not surprising to see that the network reconstruction accuracies heavily depend on signals. The stronger the signal is, the better the reconstruction accuracy can be achieved (better precision at the same recall rate or large recall rate at the same precision). Integration of genetic data improves network reconstruction accuracy (comparing red and blue lines in the same column). The improvement is obvious when the signal is weak (comparing red and blue lines in columns 1 and 2 in Fig. 8.5), which is consistent with our previous simulation results (Zhu 2007).

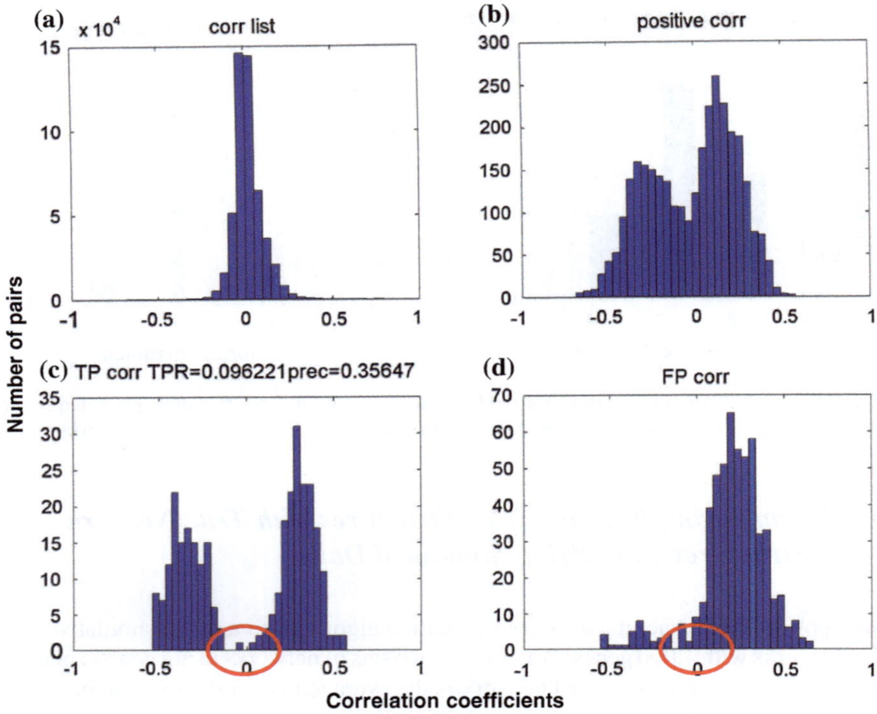

Fig. 8.7 Correlation coefficient distribution for random pairs and pairs predicted networks. **a** the correlation coefficient distribution of all possible pairs in a true underlie network; **b** the correlation coefficient distribution of all true pairs in a true underlie network; **c** the correlation coefficient distribution of all true positive pairs in a reconstructed network; **d** the correlation coefficient distribution of all true positive pairs in a reconstructed network; Some true positive links with very weak correlations (red marks) are predicted in our networks, which suggests the RIMBANet approach identifies nonlinear interactions

For two nodes that are predicted to be causally related, we tested whether they are closely related in the true network. We compared the shortest distance distribution of false positive pairs with the distribution of random pairs, shown in Fig. 8.6. Most random pairs of nodes are at least three steps away from each other (Fig. 8.6a). A small faction of false positives in our reconstructed networks is due to wrong causal direction predicted (shortest distance equal one in Fig. 8.6b). Most of false positive pairs in the reconstructed networks are indirect neighbors (two nodes share a direct neighbor) in the true underlie network structures (shortest distance equal two in Fig. 8.6b). To further dissect why false positives and false negatives are predicted, we compared pairwise correlation coefficients of all pairs (Fig. 8.7a), all true pairs (Fig. 8.7b), true positives (Fig. 8.7c), and false positives (Fig. 8.7d). Nodes that are linked in the true network structures are stronger correlated compared to nodes in random pairs (Fig. 8.7a, b). The correlation co-efficient distribution of true positives

(Fig. 8.7c) is the distribution of all true pairs (Fig. 8.7b). The stronger the correlation, the more likely it is to be recovered as expected. However, some links with pairwise correlation coefficients close to zeros (red marked region in Fig. 8.7c, d) are also predicted in our reconstructed networks.

Discussion

One of the potential drawbacks of simulation studies is that a reconstruction method tends to perform better if it shares similar assumptions with the models in simulation. In general application of RIMBNet, we discretize continuous data, then use discrete data to evaluate different structure configurations in the Bayesian network reconstruction process. The network model with discrete data can represent both linear and nonlinear relationships. However, any discretization process loses information so that the number of samples required is generally larger than linear-only models. On the other hand, the computation speed is much faster with discrete data than continuous data. Discrete Bayesian networks such as RIMBANet can be used to construct large biological networks with over 10,000 nodes (Zhu 2010). In general, it takes $1{,}000 \times 10$ CPU h (10 h for each 1,000 independent MCMC chains) to construct a network with 5,000 nodes. To apply RIMBANet to the simulated dataset of 5,000-gene networks and perform similar analysis in this chapter, at least 48 structures (24 and 24 for with and without genetic priors, respectively) need to be constructed. We did not apply our approach to the 5,000-gene network dataset as the computational requirement is out of our capacity.

In conclusion, we show that integration of genetic data into network reconstruction using our RIMBANet approach improves network reconstruction accuracy. False-positive links in reconstructed networks are not randomly distributed. More than 80 % of them connect nodes that are indirect neighbors.

References

Altshuler D, Daly MJ, Lander ES (2008) Genetic mapping in human disease. Science 322:881–888

Chen Y et al (2008) Variations in DNA elucidate molecular networks that cause disease. Nature 452:429–435

Emilsson V et al (2008) Genetics of gene expression and its effect on disease. Nature 452:423–428

Friedman N, Linial M, Nachman I, Pe'er D (2000) Using Bayesian networks to analyze expression data. J Comput Biol 7:601–620

Hsu YH et al (2010) An integration of genome-wide association study and gene expression profiling to prioritize the discovery of novel susceptibility Loci for osteoporosis-related traits. PLoS Genet 6:e1000977. doi:10.1371/journal.pgen.1000977

Leonardson AS et al (2010) The effect of food intake on gene expression in human peripheral blood. Hum Mol Genet 19:159–169. doi:10.1093/hmg/ddp476 (ddp476 [pii])

Madigan DaYJ (1995) Bayesian graphical models for discrete data. Int Stat Rev 63:215–232

Marbach D et al (2012) Wisdom of crowds for robust gene network inference. Nat Methods 9: 796–804. doi:10.1038/nmeth.2016

Pearl J (1988) Probabilistic reasoning in intelligent systems : networks of plausible inference. Morgan Kaufmann Publishers, San Mateo

Schadt EE et al (2008) Mapping the genetic architecture of gene expression in human liver. PLoS Biol 6:e107

Schwarz G (1978) Estimating the dimension of a model. Ann Stat 6:461–464

Vignes M et al (2011) Gene regulatory network reconstruction using Bayesian networks, the Dantzig selector, the Lasso and their meta-analysis. PLoS One 6:e29165. doi:10.1371/journal.pone.0029165

Witte JS (2010) Genome-wide association studies and beyond. Annu Rev Public Health 31:9–20. doi:10.1146/annurev.publhealth.012809.103723 (4 p following 20)

Zhang W, Zhu J, Schadt EE, Liu, JS (2010) A Bayesian partition method for detecting pleiotropic and epistatic eQTL modules. PLoS Comput Biol 6:e1000642. doi:10.1371/journal.pcbi.1000642

Zhong H et al (2010) Liver and adipose expression associated SNPs are enriched for association to type 2 diabetes. PLoS Genet 6:e1000932. doi:10.1371/journal.pgen.1000932

Zhu J et al (2008) Integrating large-scale functional genomic data to dissect the complexity of yeast regulatory networks. Nat Genet 40:854–861. doi:P10.1038/ng.167 (ng.167 [pii])

Zhu J et al (2010) Characterizing dynamic changes in the human blood transcriptional network. PLoS Comput Biol 6:e1000671. doi:10.1371/journal.pcbi.1000671

Zhu J et al (2012) Stitching together multiple data dimensions reveals interacting metabolomic and transcriptomic networks that modulate cell regulation. PLoS Biol 10:e1001301. doi:10.1371/journal.pbio.1001301 (PBIOLOGY-D-11-03979 [pii])

Zhu J et al (2004) An integrative genomics approach to the reconstruction of gene networks in segregating populations. Cytogenet Genome Res 105:363–374

Zhu J et al (2007) Increasing the power to detect causal associations by combining genotypic and expression data in segregating populations. PLoS Comput Biol 3:e69

Zhu J et al (2010) Characterizing dynamic changes in the human blood transcriptional network. PLoS Comput Biol 6:e1000671. doi:10.1371/journal.pcbi.1000671